STEPHEN HAWKING

STEPHEN HAWKING

A Biography

KRISTINE LARSEN

59 John Glenn Drive
Amherst, New York 14228–2119

Published 2007 by Prometheus Books

Inquiries should be addressed to
Prometheus Books
59 John Glenn Drive
Amherst, New York 14228–2119
VOICE: 716–691–0133, ext. 210
FAX: 716–691–0137
WWW.PROMETHEUSBOOKS.COM

11 10 09 5 4 3 2

Library of Congress Cataloging-in-Publication Data

Larsen, Kristine.
Stephen Hawking : a biography / Kristine Larsen. — 1st American pbk. ed.
p. cm.
Originally published: Westport, CT : Greenwood Press, an imprint of Greenwood Publishing Group, Inc., 2005
Includes bibliographical references and index.
ISBN 978–1–59102–574–0 (alk. paper)
1. Hawking, S. W. (Stephen W.) 2. Physicists—Great Britain—Biography. 3. Cosmology. 4. Big bang theory. 5. Black holes (Astronomy) 6. Space and time. I. Title.

QC16.H33L37 2007
530'.092—dc22
[B] 2007036123

Printed in the United States of America on acid-free paper

CONTENTS

Introduction 7

Timeline 11

Chapter 1. Destiny's Child: An Auspicious Birth and Eclectic Upbringing 21

Chapter 2. Scientist in Training: The Oxford Years 29

Chapter 3. Tragedy and Triumph: Deadly Disease and Dissertation 37

Chapter 4. Children and Calculations: Family Man and Theoretician 51

Chapter 5. "Stephen's Changed Everything": Black Holes Aren't Black 59

Chapter 6. Caltech and Cambridge: Exploring New Horizons 69

Chapter 7. Physics or Metaphysics?: The "No-Boundary" Proposal 81

Chapter 8. Challenges and Controversy: An Unexpected Silence and Time's Arrows 95

Chapter 9. The Best Selling Book That "No One Read": *A Brief History of Time* 107

Chapter 10. To Boldly Go: Time Travel and Television 117

Chapter 11. Plays, P-branes, and Polls: Private Lives and Public Pronouncements 129

Chapter 12. Books and Bets: *The Universe in a Nutshell* and the End of a Paradox 145

Epilogue: Stephen Hawking: Man vs. Myth 157

Afterword 161

APPENDIXES:

A: General Relativity and Cosmology 175

B: The Laws of Thermodynamics and Black Holes 183

C: Inflationary Cosmology 185

D: The AdS/CFT Correspondence 193

Glossary 195

Select Bibliography 203

Index 205

INTRODUCTION

"Hey, what's Stephen Hawking doing in this movie?"

I glanced up from my laptop, where I was furiously working on the book you now hold, and stared, bleary-eyed, at the TV. The Sci Fi Channel, which had been playing in the background, largely ignored by me, was showing *Terminal Error*, a 2002 computer-virus-plotted movie originally produced by the PAX network. What had caught the attention of my significant other was the sidekick character—a paralyzed man in an electric wheelchair, a brilliant computer scientist who could only communicate through a computerized voice operated through a keypad mounted on his wheelchair. Although the actor was obviously able-bodied, and the computer system was a fraction of the size of the real apparatus (and operated at lightning speed, as opposed to Hawking's labored clicking) there was certainly no doubt that the viewer was supposed to make the connection with the famed scientist.

How had a theoretical physicist, whose technical papers largely consisted of esoteric mathematics and shorthand diagrams only interpretable by others speaking the secret language of general relativity, become such a recognizable icon that he could be the basis of a movie character? Why had his admittedly difficult to read, popular-level book, *A Brief History of Time*, broken sales records in Britain and topped the bestsellers list in America? Who could have ever imagined that his personal life would be scrutinized by the media, hungry for any hint of a scandal? Such is the paradox of Stephen William Hawking that I have tried to unravel in this work.

As a graduate student in general relativity in the mid-1980s, I attended several conferences at which Hawking was also present, and I was fortunate to hear his lectures. These were the first years after he had lost all natural power of speech and his computerized voice was adding to the mystique of the already considerable Hawking legend. He lent an unmistakable presence to any gathering, and heads would turn as he rolled by, like Caesar in the modern-day version of a chariot, flanked by his students, attendants, and general followers. We graduate students stopped short of throwing flowers in his path, but the esteem with which we all held him was unmistakable and palpable. I distinctly remember sitting in one parallel session at a conference, with several dozen of us crammed into a small room listening to a technical talk. Hawking was there and despite the fact that he was not giving the lecture, we could not help but be aware of his presence. Suddenly a soft, rhythmic clicking could be heard. I glanced over toward the sound, saw Hawking working the mouse-like controller on his computer, and realized he was composing a question for the speaker in anticipation of the lecture's conclusion. My immediate thought was one of pity for the lecturer, and the sincere prayer that I would never find myself in the same situation!

The thought of being questioned by the brilliant mind we all knew Hawking possessed was certainly enough to strike fear in the heart of a neophyte graduate student like myself, but I was fortunate enough to experience his humorous side firsthand. At the December 1986 Texas Symposium on Relativistic Astrophysics (held in frigid and windy Chicago), I found myself invited to an impromptu private party in Hawking's hotel room. I must admit that I (and the colleague I was with) believed it to be a joke perpetrated by someone who had visited the open bar held before the banquet several times more than we had. Regardless, we went to the room in question, and when the door opened, sucked in a collective breath of shock as one of Hawking's nurses welcomed us with a request to get more glasses from our rooms, as they were running short. After a wide-eyed gape at Hawking himself sitting in the room holding court, we dutifully ran off to our rooms and fetched the glasses. I madly dialed the room of my thesis advisor, Ronald Mallett, who had gone off to bed early and knew nothing of the adventure that had fallen into my lap.

Unable to reach Ron, I continued to call him once we had returned to

Hawking's room, and when he finally answered, I tried to be as calm as I could given the situation.

"Ron, get down to room 250—now!"
"Why, Kris?"
"It's Hawking's room, and we're having a party!"

According to Einstein, the speed of light is the ultimate speed limit of the universe, but somehow Ron managed to get dressed again and arrive at the party even faster. I spent some time conversing with Hawking myself, not about physics or the meaning of life, but about family. I told him that my grandmother, who was very proud of me, only knew the names of two current scientists—Hawking and Carl Sagan. Stephen spoke about his children, including Lucy, who he then thought was going to major in history. Like his father's fears about his own career years before, Hawking worried that Lucy would have trouble finding employment after graduation. This most extraordinary of men sharing with me his all-too-common concerns about his children illustrated a side of Hawking sometimes lost in biographical studies—his ultimate humanity. Despite the fact that graduate students (myself included) considered him an object of hero-worship—no less for his triumph over his physical challenges than his mastery of the intricacies of theoretical physics—he was, at the end of the day, no different from our own parents—or ourselves—in many ways.

Hawking has stated that he does not read biographies of himself, and I am under no illusion that he will make an exception for this work. It is my sincerest wish, however, that if he did, he would find reflected in these pages the whole, complex person he truly is. Not all of what the reader will find here is flattering, but it is honest, and such is the life of any famous person if one turns over all the rocks of his or her life. The facets of his family life are balanced by his considerable contributions to theoretical physics, hopefully explained in a manner that a nontechnical reader will find illuminating. A list of Hawking's publications updated only to November 2002 lists almost 200 articles, essays, and books. His numerous public and technical lectures are omitted from that count. It is therefore impossible to discuss all of his works in a book of this level and scope, but with any luck the reader will come away with a respect for the breadth and depth of Stephen Hawking's influence on the study of black holes and the universe. Coupled with an understanding of

Hawking the man, one can understand not only what Hawking was doing "in the movie," but also why he *became* the movie, in the film version of *A Brief History of Time.*

I would like to thank Ronald Mallett and Emma Keigh for their tremendously useful comments; Brie Alsbury for the illustrations; Gary Cislak for his patience and support; and the staff of the Elihu Burritt Library, especially Emily Chasse, for their invaluable help in my research. This work is ultimately dedicated to all those who dare to soar beyond their limitations, and especially to all whom ALS has touched.

> "I would like to be thought of as a scientist who just happens to be disabled, rather than a disabled scientist." —Stephen Hawking

TIMELINE

1942 On January 8, Stephen William Hawking was born in Oxford, England, the first child of Frank and Isobel Hawking.

1943 Hawking's sister Mary was born.

1947 Hawking's sister Philippa was born.

1950 Frank Hawking became head of the Division of Parasitology at the National Institute of Medical Research, Mill Hill. The family moved to St. Albans. While Frank spent the winter in Africa, Isobel and the children spent four months with the family of poet Robert Graves in Majorca.

1953 Hawking attended St. Albans School. His spare time was spent inventing elaborate board games, which he played with close friends. Influenced by his father, he decided on a career in science.

1955 Stephen fell ill with a mysterious fever and missed the scholarship exam for Westminster School. He remained at St. Albans School much to his father's disappointment.

1956 Hawking family adopted a son, Edward.

1958 Hawking and friends constructed a simple computer dubbed LUCE.

1959 Stephen remained in St. Albans with family friends, the Humphreys, in order to study while the family traveled to India. Did well enough on college entrance exams to be accepted to Oxford for the October semester. He was several years younger than the majority of his classmates and felt isolated.

1960 After a boring first year at Oxford, Stephen became involved with the Boat Club as a coxswain. Spent an average of an hour a day on his studies; cut corners in lab in order to make more time for the river.

1962 Troubling symptoms emerged during his senior year at Oxford. Despite his lax studies, he managed a borderline first/second on his final exams. Based on an oral exam, he received a first and was accepted to Cambridge. He began graduate work under Dennis Sciama intent on working in cosmology.

1963 Jane Wilde met Stephen Hawking at a New Year's party. She was invited to his 21st birthday party soon after. Hawking spent several weeks in the hospital undergoing tests. He was finally given the devastating diagnosis of Amyotrophic Lateral Sclerosis (ALS) and the prognosis of death within two years. He returned to Cambridge with no thesis project and little hope. He and Jane began casually dating.

1964 Hawking read a preview copy of a paper by famed Cambridge cosmologist Fred Hoyle and his student Jayant Narlikar in which a new theory of gravity was proposed. At a June meeting of the Royal Society, Hoyle presented their research, which Hawking then publicly challenged. He chose the expanding universe as his general thesis topic. Jane and Stephen became engaged in October.

1965 News of Roger Penrose's singularity theorem for black holes reached Hawking through his officemate, Brandon Carter. Hawking realized he could apply the same techniques to the big bang singularity and was ready to complete his thesis. Jane and Stephen were married on July 14. In October Stephen began a research fellowship at Gonville and Caius College of Cambridge University. His first scientific paper, "On the Hoyle-Narlikar Theory of Gravitation," was published in the *Proceedings of the Royal Society.*

1966 "Singularities and the Geometry of Space-Time," Hawking's Adams Prize essay, shared top honors with an essay by Roger Penrose. Stephen officially graduated in March. Jane completed her college studies at Westfield College and began work on a PhD.

1967 Robert Hawking was born on May 28. Stephen's research fellowship was renewed for two more years. During that time he collaborated with Roger Penrose on extending singularity theorems.

1968 A joint essay by Hawking and Penrose received second prize in the Gravity Research Foundation Award.

1969 With his previous fellowship at an end, Hawking received a Fellowship for Distinction in Science. Lucy Hawking was born on November 2. Hawking's condition deteriorated to the point that he was permanently relegated to a wheelchair.

1970 After changing his research focus to black holes, Hawking proposed the existence of primordial black holes. Other important contributions included extending the proof of uniqueness theorems of black holes and devising a new definition of a black hole's horizon, which he utilized to prove that the area of a horizon of a black hole can never decrease.

1971 Essay "Black Holes" won the Gravity Research Foundation Award.

1972 At Les Houches summer school, Hawking collaborated with James Bardeen and Brandon Carter on the seminal paper of the time on black hole mechanics.

1973 Hawking's first book, *The Large Scale Structure of Space-Time,* was co-authored with George Ellis. His research interests moved to the unification of general relativity and quantum mechanics (quantum gravity). The startling discovery that black holes should radiate came near Christmas.

1974 "Hawking radiation" was officially announced in February and was met with some skepticism. "Black Hole Explosions?" appeared in *Nature.* Stephen was inducted into the prestigious Royal Society on May 2. The family spent the academic year in Pasadena, California, where Stephen held the Sherman Fairchild Distinguished Scholar visiting professorship at Caltech. Research included the application of Feynman's sum over histories path integral approach to quantum gravity. The legendary bet over whether or not Cygnus X-1 was a black hole was made with Kip Thorne in December.

1975 In January Hawking and Roger Penrose received the Eddington Medal from the Royal Astronomical Society. Stephen traveled to Rome to receive the Pius XI Gold Medal for Science from Pope Paul VI. The controversial black hole information paradox was formulated. In response to rumors that Hawking was planning to remain in California, Gonville and Caius offered him a readership, his first official post.

1976 Stephen received the Hughes Medal from the Royal Society and the Dannie Heineman Prize for Mathematical Physics from the American Physical Society. With former student Gary Gibbons, he wrote an important paper on the thermodynamic properties of de Sitter space-time.

1977 In October Hawking finally received the rank of professor, with a promotion to a special chair in gravitational physics. Jane joined St. Mark's Church choir, directed by recent widower Jonathan Hellyer Jones. They became close friends and Jonathan became a fixture in the household, aiding in the care of Stephen and the children.

1978 Hawking was given the Albert Einstein Award from the Lewis and Rosa Strauss Memorial Fund and an honorary doctorate from Oxford.

1979 Timothy Hawking was born on Easter. In November Stephen was appointed to the Lucasian Chair in Mathematics, a position once held by Sir Isaac Newton.

1980 A serious cold led to a stint in a nursing home and the beginning of part-time nursing care. On April 29, Stephen was formally inaugurated as Lucasian Professor of Mathematics delivering a controversial speech, "Is the End in Sight for Theoretical Physics?"

1981 In April Jane graduated with her PhD. At a Vatican conference in September, Stephen unveiled his preliminary work on the no-boundary proposal. He began research on the inflationary cosmology and wrote several influential papers.

1982 Stephen was invested as a Commander of the British Empire on February 23. Delivering three Morris Loeb lectures at Harvard sparked an interest in writing a popular-level book on cosmology. Gary Gibbons and Hawking hosted the Nuffield Workshop on The Very Early Universe from June 21 to July 9. In late summer Hawking collaborated with Jim Hartle at the Institute of Theoretical Physics (University of California, Santa Barbara) on the no-boundary proposal. The result was a controversial and influential paper, "Wave Function of the Universe."

1984 First draft of *A Brief History of Time* was completed.

1985 Around this time, Jane confided to Stephen that she and Jonathan had become romantically involved. Stephen received the Gold Medal from the Royal Astronomical Society. "The Arrow of Time in Cosmology" was submitted to *Physical Review D.* Shortly before publication Don Page and Raymond LaFlamme convinced him that he might be wrong about the paper's controversial result that entropy would decrease in a contracting universe. The first draft of *A Brief History of Time* was accepted by Bantam Books with many changes and revisions suggested. During a summer trip to Switzerland, a cough developed into pneumonia. Stephen was placed in a drug-induced coma and Jane was asked whether life support should be terminated.

Instead, Stephen was awakened and flew back to Cambridge to recuperate. He received a tracheotomy which permanently robbed him of his garbled speech. Stephen returned home in November to permanent, around-the-clock nursing care. Communication became possible with a computer program donated by Walt Woltosz and hardware adapted to his wheelchair by David Mason.

1986 Frank Hawking died after a long illness. Stephen resumed his travel schedule, including a trip to Rome in October where he was appointed to the Pontifical Academy of Sciences and the family was granted an audience with the Pope. At a December lecture in Chicago he formally announced his error concerning entropy and the arrow of time.

1987 The second draft of *A Brief History of Time* was completed in the spring. Stephen received the first Paul Dirac Medal from the Institute of Physics. His research turned to wormholes and their connection to possible "baby universes."

1988 Hawking and Roger Penrose were awarded the Wolf Prize in Physics. *A Brief History of Time* was released in April and became a runaway bestseller. Intense media attention was focused on the Hawking family.

1989 Thousands of free, unofficial "Stephen Hawking fan club" T-shirts were distributed by Chicago bar owners. An Honorary Doctorate in Science from the University of Cambridge was conferred upon Stephen by the Duke of Edinburgh. Queen Elizabeth designated him a Companion of Honour. In October Robert graduated from Oxford and left for postgraduate studies in Glasgow, Lucy began studies at Oxford, and Stephen announced his intention to leave Jane for his nurse, Elaine Mason.

1990 Jane and Stephen separated in February. It was kept quiet until the summer, when the press began to hound Jane. Hawking conceded his Cygnus X-1 bet to Kip Thorne.

1991 On March 5 Hawking was hit by a car and suffered a broken arm. In opposition to Kip Thorne's proposal that wormholes might theoretically be used for time travel, he developed the "chronology protection conjecture." On September 24 another bet was made with John Preskill and Kip Thorne, with the argument that the cosmic censorship conjecture was true and naked singularities were impossible.

1992 The movie version of *A Brief History of Time* premiered in Hollywood on August 14.

1993 At the home release party for *A Brief History of Time*, Hawking met Leonard Nimoy of *Star Trek*, who arranged for Hawking to have a guest appearance on *Star Trek: The Next Generation*, playing a holographic version of himself. Stephen recorded a voiceover for the Pink Floyd song "Keep Talking." *Black Holes and Baby Universes and Other Essays* was published.

1994 Hawking and Penrose held a series of debate lectures at the Isaac Newton Institute for Mathematical Sciences at Cambridge.

1995 Jane and Stephen were officially divorced in the spring. On July 8 Stephen announced his engagement to Elaine. They were married on September 16. Jane began writing her memoirs.

1996 The book version of the Hawking–Penrose debates was released. *The Illustrated A Brief History of Time* was released in November.

1997 In January Hawking created the COSMOS supercomputer consortium. The cosmic censorship bet with Preskill and Thorne was conceded on February 5, and a new more carefully worded version of the bet was made on the same day. Hawking and Thorne bet against Preskill that a solution to the black hole information paradox would not be found in the correct theory of quantum gravity. In March Lucy announced she was pregnant and that she was leaving the United States and moving back to England with her boyfriend, Alex MacKenzie Smith. Jane and Jonathan were married on July 4. Lucy's son, William, was born in the fall. *Stephen Hawking's Universe* series and book were produced by Hawking's Oxford crewmate David Filkin.

1998 Around this time, Lucy and Alex were married. Stephen and Neil Turok published a controversial paper on "open inflation." Debate ensued with friend Andrei Linde, capturing the attention of the media. Hawking delivered a "millennium lecture" at the White House on March 6. He was chosen as one of 10 influential Britons to select photographs reflecting England in the twentieth century. The 10th anniversary edition of *A Brief History of Time* was released. Hawking's research shifted to include the application of the Anti-de Sitter/Conformal Field Theory (AdS/CFT) conjecture and brane theory to various black hole and cosmological models.

1999 Stephen recorded a voiceover for an episode of *The Simpsons.* He underwent elective surgery on his larynx to prevent food from entering his lungs. He joined eleven other dignitaries in signing the Charter for the Third Millennium on Disability. He received the Naylor Prize and Lectureship in Applied Mathematics from the London Mathematical Society and the Julius Edgar Lilienfeld Prize from the American Physical Society. Jane's autobiography, *Music to Move the Stars*, was published in August. It created a furor in the press, as private details of their marriage were divulged. The year ended with an appearance by Stephen on Larry King Live.

2000 In August the play "God and Stephen Hawking" opened to mixed reviews and sharp criticism from Hawking. He provided a taped tribute for Al Gore at the Democratic National Convention. Around this time, Lucy, who had separated from her husband, learned that her son, William, had autism.

2001 Hawking drew large crowds at public lectures in Spain and India, as well as a charity lecture in Cambridge to raise money for a local school. Timothy was now a student at the University of Exeter. Hawking's public statements on controversial issues pertaining to science and society brought sharp criticism. *The Universe in a Nutshell* was released in October to general acclaim and successful sales figures. In late December Hawking lost control of his wheelchair and crashed into a wall, breaking a hip.

2002 A festschrift was held in Cambridge on January 7–11 to celebrate Stephen's 60th birthday. Stephen filed an unsuccessful complaint with the United States Trade Commission to block the publication of *The Theory of Everything* by New Millennium Press. In June he received the Aventis Book Prize for excellence in popular science writing for *The Universe in a Nutshell.* Unflattering statements by physicist Peter Higgs led to media attention.

2003 Stephen collaborated with comedian Jim Carrey to produce a skit on *Late Night with Conan O'Brien.* He received the Michelson-Morley Award from Case Western University. In December he was admitted to Addenbrook's Hospital with pneumonia.

2004 Hawking was readmitted to the hospital in February after suffering a relapse. Travel plans were cancelled. Lucy's novel *Jaded* was released in April. Media interest focused on her famous father rather than her work. The BBC drama *Hawking* aired in April to high ratings. On July 21 Hawking delivered a lecture on his solution to the black hole information paradox at the GR17 in Dublin. He conceded the bet to Preskill, but Thorne was not convinced to follow suit. Reaction was mixed as colleagues awaited a promised peer-reviewed paper. Hawking made the news in November when he called the U.S. invasion of Iraq a "warcrime" based on lies. In December he presented an award to *The Simpsons* creator Matt Groening at the British Comedy Awards.

2005 Hawking received the James Smithson Bicentennial Medal from the Smithsonian Institution on February 14. It was announced in April that he was now using an infrared device controlled by his facial muscles to operate his computer. The following month Hawking once again played himself in an episode of *The Simpsons. A Briefer History of Time,* co-authored with Leonard Mlodinow, was published in September. The same month a BBC documentary on the black hole information paradox solution was aired. A brief peer-reviewed paper on the solution appeared in *Physical Review D* in October. Reaction of colleagues continued to be muted. *God Created the Integers* was published the same month. In November, a two-week lecture tour of California and Washington was temporarily interrupted after an incident in which Hawking had to be resuscitated.

2006 On February 3 Stephen gave the Third Dennis Sciama Memorial Lecture at Oxford. Hawking announced in June that he and Lucy would be publishing a children's book, *George's Secret Key to the Universe,* in late 2007. In October it was announced that Hawking and Mlodinow's next book project, *The Grand Design,* would be published in late 2008. Elaine and Stephen filed for divorce in October. Hawking was awarded the prestigious Copley Medal by the Royal Society in November. The same month he announced in a BBC radio interview that his next goal was to go into space. Entrepreneur Richard Branson quickly announced that his Virgin Galactic company would make this a reality in the next few years.

2007 Hawking was a keynote speaker at the Royal Society in January at the occasion of the Bulletin of the Atomic Scientists' Doomsday Clock moving its hands two minutes closer to midnight. The documentary *Hawking* had its U.S. premiere on the Science Channel the same month.

CHAPTER 1

DESTINY'S CHILD

An Auspicious Birth and Eclectic Upbringing

HUMBLE BEGINNINGS

A very pregnant Isobel Hawking strolled the familiar streets of Oxford, the early January chill wrapped around her like a cloak. A few days before, she had returned to the city of her college days, fleeing the bomb-plagued home she shared with her husband Frank in Highgate, a prosperous northern suburb of London, in search of a safer place to deliver her first child. Rumors abounded as to why Oxford and Cambridge had escaped the carnage of German bombing raids. Was Hitler saving the historical college towns to be the focus of his presumed empire? Had Churchill made a similar deal with the Nazis to refrain from attacking their universities? Regardless of the reason, Oxford was now her personal sanctuary. Turned out of a hotel for fear she might deliver there, she was currently staying in the hospital, awaiting her child's arrival. With some time on her hands and a book token in her purse, she found herself in front of Blackwell's bookstore. She emerged some time later, an astronomical atlas tucked under her arm. Little did she know the prophetic nature of this purchase. Her son, Stephen, was born on January 8, 1942, precisely 300 years after the death of Galileo. Just as Galileo had forever changed our understanding of the universe and its inner workings, Stephen Hawking was destined to do the same.

Isobel Hawking was born in Glasgow, Scotland, the second of seven children born to a middle-class family doctor and his wife. Her family took on the financial burden of sending Isobel to Oxford University in the 1930s, a move made all the more remarkable by the fact that women did not commonly attend college in those days and Oxford had only started admitting women a decade before. At Oxford Isobel studied economics, philosophy, and politics, and after graduation held a variety of unsatisfying jobs, including one as an income tax inspector. Eventually she became a secretary at a medical research institute, a position clearly beneath her educational credentials.

Frank Hawking's Yorkshire farming family had once been very prosperous, but in the early years of the century its fortunes turned sour. Frank's grandfather had purchased too many farms and the family ended up in bankruptcy, saved from utter ruin only by the school that Frank's resilient grandmother opened in their home. Frank's parents were likewise financially strapped but pooled enough resources to send Frank to Oxford University, where he studied medicine and specialized in tropical diseases. When World War II broke out in 1939, Frank was conducting research in East Africa. He decided to volunteer for the military and made the arduous journey across Africa and home to England, only to be turned down because he was deemed more valuable as a researcher. Thus he took a position at a medical research institute in Hampstead, where the shy researcher met and fell in love with a secretary, fellow Oxford graduate Isobel.

The Hawking family grew during the war, with daughters Mary and Philippa born respectively 17 months and 5 years after their big brother. The closeness in their ages made Stephen and Mary natural rivals, a competition that eased only in adulthood when each carved out a unique career path—Stephen as a theoretical physicist and Mary as a doctor (dutifully following in her father's footsteps). Stephen described his youngest sister, Philippa, as "a very intense and perceptive child"[1] whose opinions he learned to respect. Stephen was a small, clumsy child who did not excel at sports or handwriting. He did not learn to read with any skill until he was eight, which he blames on his early schooling. At the Byron House School the usual methods of practice and memorization were replaced with more progressive methods which, as Stephen complained to his parents, were simply not working for him.

BEES, BOOKS, AND BETS

Frank Hawking's career took him to Africa yearly during the winter months in order to conduct firsthand research on tropical diseases. This led Mary to believe that fathers were "like migratory birds. They were there for Christmas, and then they vanished until the weather got warm."[2] He was awarded the position of head of the Division of Parasitology, and in 1950 he was transferred to the newly constructed National Institute of Medical Research in Mill Hill in the north of London. The family moved to St. Albans, a historical town famous for its cathedral and Roman ruins. In the spirit of postwar thriftiness, the family bought a rambling three-story brick house that would today be considered a fixer-upper. To the embarrassment of the Hawking children, the much-needed upgrades and repairs were never made. The house lacked central heating, and, at Frank's urging, the family and their visitors simply dressed in layers for warmth. Despite the peeling wallpaper and occasional missing glass panes that were never replaced, the house never lacked one necessity— books. Stacked two-deep on most shelves, books were the Hawkings' most obvious material possessions. Visitors to the house reported that the family never seemed to be without their noses buried in books, even at the dinner table, a custom that seemed curious, if not rude, to some of the children's friends.

"Eccentric" became the Hawkings' unofficial family label, one they did not seem to mind. The family drove a decidedly unusual car around town (a secondhand London taxi), kept bees in their basement, and made fireworks in their greenhouse. A tall and distinguished figure with white hair, Frank also had a pronounced stutter, and the children were said to speak so rapidly at times that they stumbled over their words and made up their own contractions, an accidental family dialect dubbed "Hawkingese" by friends. Frank introduced his children to surveying and astronomy, and Stephen stretched his analytical skills by devising new ways to get in and out of their house. Mary, his partner in crime, admitted that she had only ever figured out ten of his eleven escape routes. Stephen also had an imaginary house in a place called Drane, and his mother had to constantly restrain him from jumping on buses in search of this mythical house. On a trip to Kenwood House in Hampstead Heath, Stephen explained to his mother that this was his house, which had appeared to him in a dream.[3]

Stephen spent his first months in St. Albans at the High School for Girls, which took in young boys at its Michael House. One of the younger girls attending the school was Jane Wilde, who later remembered Stephen as the boy with "floppy-golden-brown hair who used to sit by the wall in the next-door classroom."[4] Although she never met the boy while he attended her school, she was later to play a pivotal role in his life. That winter, Frank's annual trip to Africa was planned for longer than usual, and Isobel decided to take the children out of school and spend four months on the island of Majorca with her longtime friend, Beryl, who was married to famed poet Robert Graves. Stephen enjoyed his time on Majorca and shared a tutor with Robert's son William.

Upon returning home, Stephen attended Radlett, a private school, for one year, then scored high enough on the eleven-plus standardized test to earn a free education at St. Albans School. By this time he was already leaning toward a career in science. He was fascinated by model trains and would take apart clocks and radios, although he was admittedly not as adept in putting them back together again. It was this constant questioning of the way things work that eventually led him to become a physicist. Hawking still explains, "I am just a child that has never grown up. I still keep asking these how and why questions. Occasionally I find an answer."[5] At the end of his first year at St. Albans School, he was third from the bottom in his class, but his teachers and friends still recognized his innate intelligence. Eventually his grades improved so that he was in the middle of his class, yet classmates called him "Einstein," another prophetic turn of events.

Stephen surrounded himself with six or seven close friends with whom he could discuss all topics that interested him, from religion to physics. He and Roger Ferneyhaugh created a series of complex board games similar to *Risk*, each of which had its own rules, game boards, and complete universes. A manufacturing game featured factories, railroads, and its own stock market, while a feudal game was so detailed that each player had a dynastic family tree. Games could last for hours or were sometimes spread over days. Friend Michael Church believed that Hawking "loved the fact that he had created the world and then created the laws that governed it."[6] When Stephen and his friends were at about the age of twelve, Basil King bet another of Hawking's friends, John McClenahan, a bag of sweets that "Stephen will turn out to be unusually capable."[7] It is no doubt that Basil has earned his candy.

Frank Hawking had always been acutely aware of the strict class structure in British society and felt that he had suffered at its tyrannical hand. He was determined that his son should reap the benefits of the finest private education possible, in preparation for what he assumed would be an illustrious career in whatever field Stephen finally chose. However, high-class schooling came with a high price tag, and the Hawkings, while frugal to an eccentric extreme, did not have the means to send Stephen without a scholarship to the prestigious Westminster School his father aspired to. Therefore at age thirteen, Stephen prepared to take the scholarship examination. Unfortunately, Stephen fell ill around this time and was unable to sit for the exam. He remained at St. Albans School and by his own admission "received as good an education or better than I would have at Westminster."[8]

The illness which kept Stephen from realizing his father's immediate educational goals proved mysterious and lingering, keeping him out of school and in bed for prolonged lengths of time. His mother later looked back upon this presumed "glandular fever"[9] as possibly the first hint of the terrible disease that would begin ravaging his body in early adulthood.

When Stephen was fourteen, his parents adopted another son named Edward. Both Stephen and those who knew his family would later recount that Edward never quite fit in with his elder siblings and their eccentric family, but Stephen believes that the addition to the family was "probably good for us. He was a rather difficult child, but one couldn't help liking him."[10]

COMPUTERS AND COLLEGES

The circle of friends that believed Stephen would be "unusually capable" remained tight during their teenage years. Stephen built model boats and airplanes with John McClenahan in John's father's workshop. He admits that his aim was "always to build working models that I could control.... Since I began my PhD, this need has been met by my research into cosmology. If you understand how the universe operates, you control it in a way."[11] The friends also explored more esoteric topics, but always with the critical eye of a scientist. Stephen became very interested in extra sensory perception (ESP), and with his friends conducted dice-rolling experiments to try and test whether or not there was any truth in its claims. A lecture on ESP recounting classic tests performed at Duke University finally convinced Hawking that it was all a fraud.[12]

Perhaps the greatest achievement of Hawking's group was the construction of a basic computer in 1958. They built it out of scavenged parts from items such as clocks and a telephone switchboard, but their Logical Uniselector Computing Engine (LUCE) soon became the talk of the town. In their last year of high school they built a more refined version, yet even this was only able to perform the simplest mathematical functions. In an unfortunate turn of events, LUCE was eventually tossed into the trash many years later by the school's new, unsuspecting head of computing.[13]

With the end of their high school career looming in front of them, it was time for Stephen and his friends to decide on colleges and, just as importantly, subjects in which to specialize. Science and mathematics had already become Stephen's passion by this time. His father, the dedicated scientific researcher, was a strong role model in his life, and Stephen enjoyed looking through the microscopes in his father's laboratory. A career in research science seemed almost a given to Stephen. But what branch of science would be his specialty? He felt biology "too inexact, too descriptive"[14] while physics was "the most fundamental of all the sciences."[15] There was also the class issue: biology was considered the realm of less capable students while physics was reserved for the intellectually elite. He was also torn by his affection for mathematics, partially thanks to a particularly inspiring teacher, Mr. Tahta. Frank Hawking was understandably disappointed that his son had no desire to follow in his medical footsteps, but he was more concerned that a course of study in mathematics would lead directly to the unemployment line. There was also the issue of where Stephen would attend college. His father was adamant in wanting Stephen to attend his alma mater, University College at Oxford University, which did not offer a major in mathematics. As a compromise, Stephen agreed to apply to University College to study physics and chemistry, with some math on the side. This plan would later haunt Stephen in graduate school, as his eventual study of the workings of the universe demanded mathematical skills he would have to learn on his own in a relative hurry.

As both his parents' families knew firsthand, an Oxford education did not come cheaply, and Stephen had to work to qualify not only for entrance into the university, but for scholarships as well. His grades in high school had been mediocre, far below his true ability, and neither entrance nor scholarships was a certainty. During Stephen's last year of high school, his father was given a long-term exchange assignment in India. Given all that rested on Stephen's performance on the upcoming exams, Stephen remained behind in England, staying with family friends, the Humphreys, while the rest of the Hawkings traveled to India for the year. Stephen was only seventeen in March 1959 at the time of the scholarship exams, and the headmaster of St. Albans thought he should wait another year. Never one to resist a challenge, Stephen took the exams, along with two St. Albans boys a year older than he, and scored quite highly on all written work. With a nearly perfect score on the physics exam and impressive interviews with, among others, the master of the college and Dr. Robert Berman, the physics tutor, there was no doubt that Stephen was headed for Oxford, scholarship in hand.

So it was that Stephen Hawking began the road to his scientific career in October 1959, at the tender age of seventeen, following in the educational footsteps of both his proud parents.

NOTES

1. Stephen Hawking, *Black Holes and Baby Universes and Other Essays* (New York: Bantam, 1993), p. 2.
2. Stephen Hawking, ed., *Stephen Hawking's A Brief History of Time: A Reader's Companion* (New York: Bantam, 1992), p. 13.
3. Ibid., p. 12.
4. Jane Hawking, *Music to Move the Stars: A Life with Stephen Hawking* (London: Pan Books, 2000), p. 9.
5. Stephen Hawking, interview by Larry King, *Larry King Live Weekend*, Cable News Network, December 25, 1999.
6. Melissa McDaniel, *Stephen Hawking: Revolutionary Physicist* (New York: Chelsea House Publications, 1994), p. 28.
7. Hawking, ed., *Stephen Hawking's A Brief History of Time: A Reader's Companion*, p. 24.
8. Morgan Strong, "Playboy Interview: Stephen Hawking," *Playboy* (April 1990): 64.
9. Hawking, ed., *Stephen Hawking's A Brief History of Time: A Reader's Companion*, p. 21.
10. Hawking, *Black Holes and Baby Universes and Other Essays*, p. 3.
11. Ibid., p. 5.
12. Dennis Overbye, "The Wizard of Space and Time," *Omni* (February 1979): 46.
13. McDaniel, *Stephen Hawking: Revolutionary Physicist*, p. 29.
14. Overbye, "The Wizard of Space and Time": 46.
15. Hawking, interview by Larry King.

CHAPTER 2

SCIENTIST IN TRAINING

The Oxford Years

THE BOAT CLUB

Oxford has a long, illustrious history, and among its more than 40 colleges and private halls, University College stands out among the rest. Founded in 1249 by Archdeacon William of Durham, University College, the oldest of Oxford's colleges, initially only admitted students of theology. Its list of former students boasts physicist Robert Boyle, writer C. S. Lewis, and former president Bill Clinton. Poet Percy Bysshe Shelley entered the college in 1810 but was expelled less than a year later for distributing a pamphlet on atheism. The Oxford calendar is very different from that of American universities, consisting of three eight-week terms running from October through December (Michaelmas), January through March (Hilary), and May through July (Trinity). This compact calendar leads to a very intense academic life—or so it seems on paper.

Stephen Hawking arrived at University College at the beginning of the Michaelmas 1959 term; at seventeen he was several years younger than the vast majority of his fellow students. Not only had he taken the entrance exams a year early (passing his trial run, as it were), but most of his class had completed mandatory military service before entering college. The draft had been abolished just in time to save Hawking from the experience. This age difference isolated Hawking from the start. In addition, the college students of his generation were generally disillusioned with the mainstream culture of the world around them. They were bored with life, disdained the materialism of the "Establishment," and immersed themselves in an Oxford subculture where to work hard for grades was considered the greatest sin. Without the comforting familiarity of his high school friends, and among older strangers in a culture of boredom and detachment, Hawking found himself feeling lonely and decidedly unhappy for his first of three years at Oxford.

Even the physics program itself seemed crafted in such a manner as to segregate him. There were only four physics students entering University College that year: Hawking, Gordon Berry, Richard Bryan, and Derek Powney. The physicists spent much of their time together, both inside and outside of the classroom, and Hawking became close friends with Gordon, his tutorial partner. In their second year of college, they became part of a University tradition that was to finally change Hawking's outlook on college life.

Then, as today, college athletics were fiercely competitive and created passionate rivalries. No sport was taken more seriously at Oxford than rowing. Not only was there a long-standing yearly rivalry with Cambridge, Oxford's intellectual competitor, but there were also internal competitions where each of the individual colleges competed for bragging rights. University College had fielded rowing teams since 1827 and was an active participant in the "Eight's Week" event held on the Isis (as the Thames River is affectionately called) near the end of Trinity (summer) term. The Boat Club, as it was called, frequently recruited muscular men for rowers (eight per boat), while also spying for smaller, lighter men to act as the coxswain (or cox), the person who sits in the front of the boat facing the rowers and barks out orders while steering. Gordon Berry and Stephen Hawking were both recruited as coxswains.

The Boat Club was as much a social institution as an athletic one, and Stephen finally found himself part of the "in" crowd. Just as he had enjoyed controlling models and make-believe board game universes as a child, he reveled in his maritime reign. Norman Dix, the College Boatsman, described him as "rather an adventurous type; you never knew quite what he was going to do when he went out with the crew."[1] He would sometimes return with broken oars and damage to the boat from steering in tight places. Dix attributed that partially to the fact that Hawking was probably doing physics in his head while he coxed. In today's language, he would be said to have been multitasking. Friends remember Hawking in his second and third years at Oxford as "lively, buoyant, and adaptable. He wore his hair long, was famous for his wit, and liked classical music and science fiction."[2] This was a far cry from the lonesome seventeen-year-old who had entered the university the year before.

COSMOLOGY CALLS

Stephen found no challenge in the physics curriculum. He called it "ridiculously easy. One could get through without going to any lectures, just by going to one or two tutorials a week. You didn't need to remember many facts, just a few equations."[3] Perhaps Stephen found it easy going, but his classmates fervently disagreed. Derek Powney recounted how the four of them were assigned thirteen homework problems in an Electricity and Magnetism course, with the encouragement from their tutor, Robert Berman, to just do as many as they could. At the end of a week, Richard and Derek had only finished one and a half, Gordon one, and Stephen had not even begun the assignment. The next day he skipped his lectures and completed ten of the problems on his own, before lunch. Derek admitted that at that point he and the others "realized that it was not just that we weren't in the same street, we weren't on the same planet."[4]

Tutor Robert Berman called Hawking the most brilliant student he had ever had. He noted that Hawking didn't bother buying many of the textbooks and never took notes. However, he also explained that despite his genius, Hawking managed to come down to earth and be "one of the boys. If you didn't know about his physics and to some extent his mathematical ability, he wouldn't have told you.... He was very popular."[5]

That popularity was, of course, connected to his rowing. Stephen had to balance his time between his studies, such as they were, and his time on the river. Rowing demanded many hours of practice, six afternoons a week, which cut into the time he was supposed to spend doing experiments in his laboratory course. According to Gordon Berry, he and Stephen cut serious corners in taking data, faking their way through parts of the experiments by using creative analysis to write their lab reports.

The course of study at Oxford was three years, and in his final year Stephen had to begin making plans for graduate study. He knew that physics would continue to be his field, but he had to decide on an area of specialization for his PhD. Once, he had taken a summer course at the Royal College

Observatory with Sir Richard Woolley, the Astronomer Royal, taking measurements of double stars. He had found the work tedious and uninspiring and knew that observational astronomy was most definitely not of interest to him.[6]

Theoretical physics in the early 1960s was progressing on two important fronts—cosmology, the study of the universe on the grandest scale, and particle physics, the study of the universe on the tiniest scale. Both were ultimately derived from the work of Albert Einstein: cosmology was based on the general theory of relativity, his radical rewriting of our understanding of gravity, while particle physics evolved from quantum mechanics, the rules that govern the universe on the atomic and subatomic scale. Einstein had been one of the somewhat reluctant forefathers of quantum mechanics but was never fond of its philosophically fuzzy predictions. Hawking felt that particle physics "seemed like botany. There were all these particles, but no theory."[7] Cosmology, on the other hand, was based on a "well-defined theory."[8]

Having decided on a topic, the matter of deciding on a university still faced him. Oxford did not have any cosmologists on staff. Cambridge featured the brilliant, if not eccentric, Fred Hoyle. A noted science fiction writer and science popularizer who had become somewhat of a household name through a series of radio talks which influenced a generation of science students, Hoyle was also a scientist of international reputation. He had made important strides in the understanding of the chemical evolution of stars and was one of the founders of the steady state model of cosmology, the archrival of the big bang model. In fact, Hoyle coined the name "big bang" to poke fun at the theory he thought preposterous.

The concept of the big bang owes its genesis to the work of American astronomer Edwin Hubble. He had discovered in the 1920s that the galaxies were apparently moving away from each other, and the farther away a galaxy was, the faster it was receding. Today, the relationship between a galaxy's distance and its recessional velocity is called Hubble's law. Belgian priest and astrophysicist Georges Lemaître applied Einstein's equations of general relativity to Hubble's discovery and explained this as an overall expansion of the universe, which had originated from a "giant 'primordial atom' which exploded because of violent radioactive decay processes."[9] Russian physicist George Gamow (a noted popularizer of science who had immigrated to the United States) and his colleagues modified Lemaître's ideas and crafted the model which became known as the big bang. It predicted that after being created in an initial hot, dense state, the universe would become less densely packed and cooler as it evolved, as the galaxies were carried farther away from each other by the overall expansion of the very fabric of the universe.

The steady state model countered by saying that the empty space between the moving galaxies was constantly being filled with new material, which was being spontaneously created from nothing, and therefore the overall appearance of the universe would always remain constant—the density of the universe would never change. Fred Hoyle tweaked Einstein's equations to permit such creation. Whereas the big bang predicted that the universe had a definite beginning and changed over time as it evolved, the steady state claimed that the universe was eternal and unchanging in the sense that it always looked the same, on average, in all places and at all times—it was in a steady state of being. The 1950s were a time of vigorous, and sometimes bitter, debate between the two groups of scientists. It was not until the early 1960s that a glimmer of resolution regarding the debate appeared on the horizon—in favor of the big bang.

"IF I GET A FIRST . . ."

This was the exciting scientific realm in which Hawking wished to become embroiled. There was only one, rather significant, barrier in his way. He was accepted into Cambridge, but with the condition that he receive a "first" on his final exams, the equivalent of high honors. This was not a given, as Hawking had not kept up with his studies as he should have over the previous three years, later estimating that he had only done the equivalent of an hour's work per day. "I'm not proud of that," he admitted. "I'm just describing the attitude at the time, which I shared with fellow students. . . ."[10] Fortunately there was a choice of problems on the exams from which Hawking could select, and he counted on the fact that there would be enough theoretical questions, which he could easily handle. After the exams, three of the four physics students were depressed, believing they had not achieved the scores they had wanted. Only Gordon was upbeat, believing he had scored well enough for a first.

In one of his more practical moments, Hawking had created a backup plan and applied for a job with the Ministry of Works. However, on the morning of the Civil Service exams, just after his finals, he overslept (as usual), missing the tests. In the end it turned out to be just as well, as all four of the physicists had miscalculated their scores. All but Gordon had actually scored better than they had anticipated: Derek and Gordon received seconds, Richard a third, and Stephen a borderline grade between a first and a second. At the interview to determine his final grade, Stephen was asked what his plans were for after graduation. With his usual Hawking aplomb, he matter-of-factly stated, "If I get a first, I shall go to Cambridge. If I receive a second, I will remain at Oxford. So I expect that you will give me a first."[11] Hawking received his first and made plans to attend Cambridge in the fall.

The future seemed bright for the young man from St. Albans, but there was a dark storm cloud whose presence was becoming harder to ignore. In his last year at Oxford, Hawking noticed that "I seemed to be getting more clumsy, and I fell over once or twice for no apparent reason."[12] He also found he had difficulty rowing a sculling boat. He hid his puzzling symptoms from his family, but his friends witnessed an event that could not be ignored so easily. Near the end of his last term, he fell down a flight of stairs, landing on his head. It is reported that he may have blacked out briefiy, and he suffered a temporary loss of memory. It took several hours before he recovered all his memory, with the patient help of his alarmed physics friends. Concerned about his physical and mental health, he took a Mensa intelligence test and proved to himself and his friends that he was still a genius.[13] A trip to the doctor turned up no physical damage, and no hint of the true underlying cause of his increasing difficulties.

He took a summer trip to Persia (now Iran) with a friend, and became seriously ill while there. At the time he blamed it on a stomach ailment, or possibly a reaction to a vaccine received before the trip, but he returned visibly weaker than before. The truth of the matter was more ominous still but would not be uncovered until after his first term at Cambridge.

NOTES

1. Stephen Hawking, ed., *Stephen Hawking's A Brief History of Time: A Reader's Companion* (New York: Bantam, 1992), p. 39.

2. Kitty Ferguson, *Stephen Hawking: Quest for a Theory of Everything* (New York: Franklin Watts, 1991), p. 36.

3. Stephen Hawking, *Black Holes and Baby Universes and Other Essays* (New York: Bantam Books, 1993), p. 165.

4. Hawking, ed., *Stephen Hawking's A Brief History of Time: A Reader's Companion*, p. 36.

5. Michael Harwood, "The Universe and Dr. Hawking," *New York Times Magazine* (January 23, 1983): 57.

6. John Boslough, *Stephen Hawking's Universe* (New York: Quill/William Morrow, 1985), p. 23.

7. Alan Lightman and Roberta Brawer, *Origins: The Lives and Worlds of Modern Cosmologists* (Cambridge: Harvard University Press, 1990), p. 396.

8. Hawking, ed., *Stephen Hawking's A Brief History of Time: A Reader's Companion*, p. 59.

9. George Gamow, "Modern Cosmology," in *The New Astronomy* (New York: Simon and Schuster, 1955), p. 23.

10. Morgan Strong, "Playboy Interview: Stephen Hawking," *Playboy* (April 1990): 66.

11. Boslough, *Stephen Hawking's Universe*, p. 23.

12. Gregg J. Donaldson, "The Man Behind the Scientist," *Tapping Technology*, May 1999, www.mdtap.org/tt/1999.05/1-art.html (accessed November 9, 2001).

13. Hawking, ed., *Stephen Hawking's A Brief History of Time: A Reader's Companion*, p. 44.

CHAPTER 3

TRAGEDY AND TRIUMPH

Deadly Disease and Dissertation

STEADY STATES

In the preface to his popular-level book *The Unity of the Universe*, Cambridge physicist Dennis Sciama describes cosmology circa 1961 as "a highly controversial subject, which contains little or no agreed body of doctrine."[1] This accurately reflected the bitter debate that raged on during that time over the big bang and steady state models. Did the universe have a beginning? If so, did this mean that God had to be invoked as an ultimate explanation? Could one even trust the laws of physics as we know them today to hold all the way back in time if the universe began in the hot, dense state predicted by Gamow and his colleagues?

Sciama was in the thick of the debate, and given the fact that his book was dedicated to Bondi, Gold, and Hoyle, the "holy trinity" of the steady state model, there was no doubt which theory he favored. But this was not a case of blind loyalty. Sciama had held a research fellowship at Cambridge from 1949 until he received his PhD in 1953. He had witnessed the birth of the steady state, and knew the three authors of the theory, each of whom had spent some time at Cambridge. Bondi, Gold, and Hoyle, "all colorful personalities, excellent speakers, and forceful writers, applied themselves zestfully to the task of convincing the world of the rightness of the steady-state theory."[2] Sciama was convinced to the point that he found the theory attractive, and believed that the spontaneous creation of matter was "even less of a thing to introduce than the creation of a whole universe at one go."[3] But it wasn't just the theory itself that Sciama found compelling, but the personalities involved. They were young and brash, scientific rebels who flew in the face of establishment. Sciama found them "an exciting influence for a younger person like myself."[4]

Thus, in 1962 the Cambridge Department of Applied Mathematics and Theoretical Physics (or DAMTP) was stacked in favor of the steady state model. Sciama was a lecturer in the department, and Fred Hoyle was the Plumian Professor of Astronomy and Experimental Philosophy. However, irrevocable change was on the horizon. Martin Ryle, who had worked on radar during World War II and later won the Nobel Prize in Physics, led the radio astronomy group at Cambridge. By 1962 he had found that the universe did not seem to be as uniform as the steady state suggested, through what was known as radio counts. This involves counting the number of astronomical radio sources in different directions in the universe and at different distances from Earth. He had found that many of these radio sources (which we now know to be galaxies with voracious black holes in their cores) seemed more common the farther out in space his instruments probed. Because it takes the light some time to travel through space and reach Earth, the farther out one looks in space, the older the light is that is reaching us. Therefore, these radio sources were more common the farther back in time one looked. In response, Hoyle and one of his graduate students, Jayant Narlikar, began making adjustments in the steady state model to explain the apparent contradiction.

Change and uncertainty were also afoot in St. Albans. In the summer of 1962, a group of schoolgirls had just finished their final exams and were enjoying the freedom of summer. As they crossed a street, a young man with unruly hair and an "awkward gait" crossed in the other direction, seemingly lost in his own thoughts. Diana King announced to her friends that the eccentric but brilliant youth was Stephen Hawking, a friend of her brother, Basil. She then bragged, "I've been out with him actually."[5] Jane Wilde, one of her companions, noted the youth (a former elementary school classmate) with curiosity. The daughter of a civil servant, Jane was a serious and shy young woman, and although dedicated to her studies, she was not certain she would succeed in getting into Oxford or Cambridge. While her friends were leaving school that summer, she planned to remain for the fall term as Head Girl while she applied for college. To her disappointment, and that of her father (who had hoped for a Cambridge education for his child since she was very young), Cambridge was not to be in her future, and neither was Oxford. Instead, she was accepted to Westfield College, a ladies' art college in London, where she planned to study Spanish and French.

The fall term at Cambridge found Stephen Hawking dealing with his own disappointments and challenges. Cosmology was not highly regarded by much of the scientific community at that time: it was believed to be merely speculative, due to the small amount of observational evidence and the lack of doctrine that Sciama had earlier lamented. Hawking understood the precarious position of his chosen area of study yet found something positive in it. He found cosmology and the study of Einstein's general theory of relativity to be

> neglected fields that were ripe for development at that time. Unlike elementary particles, there was a well-defined theory . . . thought to be impossibly difficult. People were so pleased to find any solution to the field equations; they didn't ask what physical significance, if any, it had.[6]

He had further been inspired by a summer course that he had taken with Jayant Narlikar and looked forward to studying under the legendary Fred Hoyle.

But this was not to be. Hoyle already had a sufficient number of graduate students working under him, so Hawking was assigned to Dennis Sciama. Hawking had never heard of Sciama and was understandably disappointed. However, he found that this original setback turned out to be for the best. Hoyle had a joint appointment with major observatories in America and was away from campus for significant periods of time. Sciama, on the other hand, was a homebody, and was always available for discussion and guidance. He was known to be warm and enthusiastic, putting the needs of his students clearly before his own. In fact, it was because of this relegation of "his own personal research and career to second place, after those of his students, [that] he was never promoted to the august position of 'Professor' at Cambridge. . . . It was his students, far more than he, who reaped the rewards and the kudos."[7] Hawking did not always concur with his advisor's scientific viewpoints, but he always found him stimulating. For his part, Sciama had a different opinion on the situation. Noting that Cambridge attracted the very best students, he felt that his role was to give them the background and structure they needed, and then just "sit back and let [them] go."[8]

Hawking initially proved to be the exception to Sciama's rule. He struggled through the first year or so of graduate school, partially due to his weak mathematical background compared to some of the other students, as well as the lack of an obvious problem on which he wanted to work. Sciama suggested he might want to work on astrophysics, but Hawking was determined to work on cosmology and general relativity, regardless of his difficulties. He read up on general relativity on his own, and with fellow graduate students routinely traveled to lectures at King's College in London, where Hermann Bondi had set up a program in general relativity. One final roadblock threatened his progress—those mysterious symptoms, which had first appeared at Oxford, were now becoming more difficult to ignore. Sciama could chalk up Hawking's slurred speech to an impediment, but it would be impossible to fool those who knew the graduate student well.

A DEVASTATING DIAGNOSIS

During the Christmas break from college, Stephen's deteriorating condition and escalating clumsiness could not be hidden from friends and family. During an ice skating outing with his mother, he fell on the ice and was unable to get up. Alarmed, his mother later took him to a local café and coerced her son to admit to the increased difficulties he had been having. At his father's insistence, Stephen visited his family doctor, who could not diagnose the problem and made a referral to a specialist for after the holidays.

With this uncertainty hanging over his head, Stephen made the best of his break from school, attending a New Year's Day party thrown by his long-time friend, Basil King, and Basil's sister, Diana. Also in attendance was the shy Jane Wilde. She listened to his tales of Cambridge with rapt curiosity, drawn by "his sense of humor and his independent personality."[9] She thought him a kindred spirit, sensing a painful shyness behind the gregarious exterior. At the end of the party they exchanged names and addresses, but Jane did not expect anything more to come of it. To her delighted surprise, she received an invitation a few days later to a party on January 8. What the invitation did not explain was that the party was in celebration of Stephen's 21st birthday (Diana King told her this in private). At this particular event, Stephen was surrounded by his college friends, and Jane felt keenly out of place with this older crowd—after all, she had not even begun her first year as an undergraduate, and

Stephen and his friends were graduate students. In the weeks after the party, she turned her attention to commuting to a secretarial course in London, where she honed her shorthand and typing skills.

Stephen was facing a much greater challenge. Shortly after his birthday, he entered St. Bartholomew's Hospital (known as St. Bart's), where his sister Mary was training to become a doctor, and spent two long weeks undergoing a barrage of unpleasant medical tests. In the end he was given the devastating news—he was suffering from Amyotrophic Lateral Sclerosis (ALS), also known as Motor Neurone Disease in Britain. Commonly called Lou Gehrig's disease, after the Yankees baseball player who died of the disease in 1941, the ailment is an unexplained, incurable, debilitating deterioration of the ability to control voluntary muscles in the body. Motor neurons, nerve cells of the brain and spinal cord, gradually die, as well as the nerve fibers that connect them to the muscles, leaving muscles unable to move or function.[10] The vague symptoms that Hawking had been suffering for over a year, including slurred speech, tripping, and general clumsiness, are classic symptoms of the disease's early stages.

Those living with the disease face an uncertain future. Walking becomes increasingly difficult, leading to the need for a wheelchair. Arms and hands become weaker, so even simple tasks such as eating and writing become a challenge. Speaking and swallowing also become labored, and finally even breathing becomes a battle. Patients in advanced stages may be placed on a respirator, and death from pneumonia is a serious threat. However, the involuntary muscles are left untouched (such as the heart, the muscles involved in digestion and waste elimination, and the sexual organs) and, most importantly, the patient's mind is not affected. With the help of medical technology, the patient can carry on with a productive life, but the prognosis is grim and patients often die within two years. However, in some cases, patients have been known to live for decades after the diagnosis, with younger patients living the longest, and male patients living longer as well.

Hawking's doctors clearly thought his life expectancy would be toward the short end of the spectrum, and they painted a bleak future for the stunned graduate student and his family. Stephen was understandably shocked. "Why should it happen to me?" he asked himself. "Why should I be cut off like this?" However, while he was lying there, feeling sorry for his apparent fate, he watched a young boy die of leukemia in the bed across from him and realized there were people with a harder fate than his. For many years, he remembered that boy whenever he fell into a bout of self-pity. His dreams were also disturbed for some time after his diagnosis. Admittedly, he had been bored and unchallenged in his life, his time at Oxford being a prime example. Now he dreamed that he was going to be executed and suddenly "realized that if I were reprieved, there were a lot of worthwhile things I could do." He also dreamed several times about sacrificing himself for others, which he rationalized by "if I were going to die anyway, it might do some good."[11]

Stephen found some solace in the operatic works of Wagner, which "suited the dark and apocalyptic mood I was in."[12] As for his studies, the doctors encouraged him to continue, but he thought it pointless, as he would probably not live long enough to finish his degree. With his doctors admitting there was little they could do aside from encouraging him to distract himself with his studies, Hawking turned to his father, the expert in tropical diseases, for advice and hope. Frank Hawking researched every possible lead, contacting experts in related diseases, no matter how obscure. He even investigated Kuru, a rare illness with similar symptoms but spread by cannibalism, in the hope of finding some small possibility of treatment.[13] With seemingly little to look forward to, Hawking did as he was advised and returned to Cambridge after the holiday break to continue his graduate work, which seemed to be going nowhere.

Jane Wilde looked forward to weekends, when she could spend time with her friends and forget about her secretarial studies. Toward the end of Stephen's stay in the hospital she ran into Diana King, now a nursing student, and was shocked to hear the troubling news about the flamboyant young man she had found so intriguing. Jane's resulting sorrowed silence was noticed by her mother, who suggested Jane pray for Stephen, as there was little else that could be done.

HOPE FOR THE FUTURE

An unexpected meeting at the train station about a week later was to change Jane's life forever. There was Stephen, returning to Cambridge on the same train she was taking to her secretarial classes. To her infinite surprise, he was not only pleased to see her but apparently in a generally cheerful mood, and unwilling to talk about his disease. They spent the train ride lost in conversation with each other and at the end made plans for their first official date—dinner at a fancy Italian restaurant and tickets for the theater.

That auspicious first date led to an invitation to the Cambridge formal event known as the May Ball (actually held in June), still some months away. Therefore Jane did not see Stephen for several months, taking various secretarial jobs to save money for a summer trip to Spain. When Stephen arrived to take her to Cambridge for the ball, the obvious deterioration in his condition stunned Jane, but she enjoyed their time together nonetheless. She became painfully aware that their budding relationship was likely doomed to be short-lived and would probably lead to a broken heart.

She contemplated this during her summer trip to Spain, the miles apart painfully felt. She "longed to have someone with whom to share my experiences. Moreover, I realized that the person I most wanted to share them with was Stephen."[14] Doubts crept into her mind—could she help him to "find even brief happiness? I doubted whether I was up to the task."[15] Returning home before the fall term, Jane found that Stephen had already returned to Cambridge and, according to his mother, was not physically well. Disheartened, she turned her energies to preparing for her first term as a student at Westfield. She did not hear from Stephen until November, when he asked if she would like to go to the opera with him in London—Wagner's "The Flying Dutchman."

After a long absence from Stephen, Jane was highly aware of the decline in his physical condition. His gait was now uncertain and lurching, and he was unable to walk for long distances. After their date, Jane felt a burning need to learn more about Stephen's condition, but her research attempts came up empty. She grew philosophical about her lack of knowledge, thinking that ignorance might be best. After all, no one had a guaranteed length of life, even those believing themselves to be in perfect health.[16] She briefly sought support in the college Christian Union, a response to Stephen's unabashed atheism. She worried that his lack of faith would "destroy us both.... I needed to cling to whatever rays of hope I could find and maintain sufficient faith for the two of us if any good were to come of our sad plight."[17]

Jane was not the only one to view Stephen's steep physical decline with alarm. After being told that his son had two years to live, Frank Hawking pleaded with Dennis Sciama to hurry along Stephen's thesis, worrying that he would not live long enough to complete it any other way. While Sciama was aware of Stephen's intelligence and potential, he refused to make any special exceptions and would not rush Stephen's thesis. This left Stephen floundering without a main thesis topic, as he had not yet found a suitable problem to tackle.

Christmas break was a welcome respite from the disappointments of Cambridge, and Jane attended the opera with Stephen and his father. She discovered that opera had become a staple in the Hawking household, and over the next few months she and Stephen attended many such performances when he came to London for general relativity seminars. Jane, in turn, visited Cambridge on many weekends, but their relationship was not without its problems. Stephen refused to discuss his disease, and understandably did not consider their relationship in the long term. Finally, his doctors essentially washed their hands of him, unable to halt the relentless slide in his physical condition and equally unable to offer him any hope. With their relationship on unsure footing, Jane left in April for a term in Spain. She repeatedly wrote to Stephen while away, but he never answered.

Although his personal life was currently under a dark cloud, Stephen found a glimmer of interest in that spring 1963 term. Jayant Narlikar, Hoyle's graduate student who had sparked Hawking's interest in cosmology in a previous summer class, had an office adjacent to Stephen's, and they freely discussed physics. Hawking was fascinated with the modifications to general relativity that Hoyle and Narlikar were making in order to reconcile the recent observations of the cosmos, such as Ryle's aforementioned radio counts, with the steady state model. Hawking eagerly studied Narlikar's calculations and made some of his own.

In June, Hoyle gave a talk at the prestigious Royal Society, unveiling his recent work with Narlikar before it was scheduled to be published. In the customary question and answer period following the talk, Hawking openly challenged one of Hoyle's results. Hoyle pressed Hawking to explain how he could know the result was wrong, and Hawking replied confidently that he had worked it out on his own. The crowd assumed Hawking had done it in his head on the spot, not knowing that he and Narlikar had been discussing the results for some time. People began taking note of the previously unknown graduate student, and Hawking found some focus for his research—he would study various properties of the expanding universe. It still wasn't specific enough for a complete thesis, but it was a beginning.

Later in the summer of 1963, Stephen set off for a famed Wagner festival in Bayreuth, Germany, with his younger sister, Philippa (who had been accepted by Cambridge to study Chinese), while Jane toured Europe with her family. To Jane's delight, a postcard from Stephen was awaiting her when she arrived in Venice. When both families returned to St. Albans at the end of the summer, Jane and Stephen had a warm reunion, and their relationship blossomed in earnest. His physical condition had stabilized, although he now used a cane, and Stephen seemed to find joy and hope in the promise of their future together. He proposed to Jane in October, at the beginning of the fall term. She happily accepted, eagerly giving up her career plans in diplomacy.

In the convention of the times, Stephen formally asked George Wilde for his daughter's hand in marriage, a request which was granted under one condition—that Jane would complete her college education. Stephen also addressed his future father-in-law's concerns that he would be an unreasonable burden to Jane, promising that he would not make demands of Jane that she could not "reasonably accomplish."[18] Many years later, Jane admitted that if her own daughter brought home someone like Stephen in a similar situation, she did not know how she would react.[19] Frank Hawking warned Jane that Stephen did not have long to live, and that they should have children as soon as possible, if that was their wish, assuring her that Stephen's condition would not be passed on to them.[20]

With the optimism and perhaps naiveté of her young age, Jane set her mind to marrying Stephen, regardless of the challenges and obstacles and even the grim expected outcome. She willingly accepted the responsibilities of taking on all the household duties and caring for Stephen, a task that was sure to become harder as the months, and perhaps years, passed. In return, all she expected of her husband-to-be was that he would love her in return and encourage her in whatever personal interests she might have time to pursue.[21] For his part, Stephen, freely and often, admitted that the engagement "changed my life. It gave me something to live for."[22] Jane later reminisced about those days of idealistic innocence, explaining that

> we had this very strong sense at the time that our generation lived anyway under this most awful nuclear cloud—that with a four-minute warning the world itself could likely end. That made us feel above all that we had to do our bit, that we had to follow an idealistic course in life. That may seem naïve now, but that was exactly the spirit in which Stephen and I set out in the Sixties—to make the most of whatever gifts were given us.[23]

A SINGULAR IDEA

In addition to Stephen's illness, there were two very concrete obstacles to their plans. First, Westfield did not normally allow its undergraduates to be married.

Jane managed to obtain the necessary special permission, since Stephen might not live long enough to marry her if the wedding date was delayed until after her graduation. There was, however, a price to be paid—Jane would have to move off campus into a private room or apartment during the week, while Stephen would remain in Cambridge and move into new housing they would only share on the weekends.

Second, Stephen would have to get a job in order to support them. To do this, he would have to find a thesis topic, finish his research in a timely manner, and apply for a position, ideally a research fellowship. This would mean that, for the first time in his life, Stephen would have to actually work hard at his studies. Hawking later admitted that to his surprise, "I found I liked it. Maybe it is not really fair to call it work."[24] Hawking began by writing up his corrections to Hoyle and Narlikar's theory of gravity and submitted it to the Royal Society for publication.[25] However, he still needed a specific problem to study for the end of his thesis. One was to soon fall into his lap, but not from the study of the steady state model.

Cosmology was certainly not the only application that physicists had found for gravitational studies. In the 1700s, Reverend John Mitchell had speculated about strange stars whose gravitational field was so intense that not even light could escape from them. In 1916 Karl Schwarzschild used Einstein's equations of general relativity to represent the gravitational field of an object in which all its mass was crushed into a single point of unimaginably high density—a *singularity*. Such a mystifying object was predicted to be enveloped within a "boundary of no return," from which no object, information, or beam of light could escape. This was termed the *Schwarzschild radius*, or *event horizon* (as events inside this boundary could not be observed by those outside of it), and the fact that light itself is trapped gives the object its modern (from the late 1960s) name of black hole.

Scientists had no reason to believe such strange, theoretical objects would exist in nature until the 1930s, when Subrahmanyan Chandrasekhar, J. Robert Oppenheimer, and others began unraveling the unusual properties of stellar corpses. The life of a star is essentially a carefully choreographed balancing act between the inward pull of the gravitational force that created the star out of its initial cloud of gas and dust (or nebula) and the outward flow of energy created by nuclear fusion in the core of the star. When a star ceases to produce energy, gravity takes over, and the star can collapse indefinitely unless some other force steps in to counteract gravity. It was discovered that for stellar corpses heavier than about three times the mass of the sun (or three solar masses) there is no force known in the universe that can stop the collapse of the star to a singularity.

In the 1960s a flurry of theoretical work began on these still hypothetical objects (dubbed "collapsed stars" in the West and "frozen stars" by Russian physicists). One important question that physicists and mathematicians tried to tackle was the necessity of the singularity. Since the very laws of nature appeared to break down at the infinitely dense heart of the collapsed star, might it be possible for nature to somehow avoid this unpleasant state of being? Roger Penrose was to soon demonstrate that the answer was a resounding no.

Penrose's interest in mathematics and science had been fanned by the radio broadcasts of Fred Hoyle in the late 1940s. In the 1950s he attended Cambridge University as a research student in mathematics and came under the encouragement of Dennis Sciama. Penrose recounted fondly that he learned "an awful lot from Dennis—not just cosmology but physics generally, and a kind of excitement and enthusiasm for the subject that was very important to me."[26] After completing his PhD he eventually joined the staff of Birbeck College in London. Penrose applied his mathematical ability and his interest in cosmology and physics to a study of the singularity problem via a new approach. In January 1965 he announced at a seminar at Kings College, London, that singularities were a general feature of gravitational collapse—the first of the so-called singularity theorems.

Hawking heard about the ground breaking results from his officemate, Brandon Carter, and immediately realized that he had found the problem he had been looking for. He went to Sciama's office and announced that "similar arguments could be applied to the expansion of the universe. In this case, I could prove there were singularities where space-time had a beginning"—namely in the big bang.[27] Hawking did just that and, working alone and with George Ellis, he was able to apply Penrose's techniques to various models of the universe. Hawking's first singularity theorem became the basis for the final chapter of his thesis and led to a series of scientific publications in the years that followed.[28]

While he began working on his newfound ideas, Hawking was also tending to the practical problem of securing employment after graduation. In February he completed an application for a research fellowship at Gonville and Caius College (part of Cambridge). Sciama wrote him a letter of reference, but he needed a second, so after a lecture Hawking asked Hermann Bondi if he would provide one. Bondi was one of the founders of the steady state model, and it was he who had submitted Hawking's paper on Hoyle and Narlikar's theory of gravity to the Royal Society the previous fall. Bondi agreed, but when the fellowship committee asked for the reference, Bondi somehow did not remember Hawking. Sciama cleared up the embarrassing misunderstanding, and Hawking was given the position, to begin in October 1965.

That spring Hawking also applied for the privately funded Gravity Prize, but his application missed the postmark deadline and he only received a commendation prize, which nonetheless gave him 100 pounds toward the cost of his upcoming wedding.[29] At the threshold of his professional career, Hawking had the opportunity to attend his first international conference on general relativity (held in London) where he began networking with colleagues from around the world (including Kip Thorne from the California Institute of Technology [Caltech], who would become a valued friend).

Not only were collapsed stars a hot topic of discussion, but also the battle between the big bang and steady state models was reaching its conclusion. Martin Ryle and others had continued to find that the distribution of radio galaxies and their relatives, quasars, did not match that predicted by a steady state universe, and, more importantly, a faint echo of energy predicted by George Gamow as a remnant of the big bang (now called the *cosmic background radiation*) had been accidentally discovered by Arno Penzias and Robert Wilson, engineers for Bell Labs in New Jersey in 1964 (a discovery which would eventually win them the Nobel Prize in physics). By the summer of 1965, it was becoming nearly impossible to support the steady state model, and Dennis Sciama, for one, gave up trying. Hawking noted that it was therefore "just as well I hadn't been a student of Hoyle because I would have had to have defended the Steady State."[30] The big bang had proven a fertile playground for Hawking, and he was just as happy to see it supported by observational evidence.

As a capstone of this most promising school year, Jane and Stephen were formally wed in a civil ceremony on July 14, 1965, followed by a religious service the following day at the chapel of Trinity Hall, Cambridge. After a brief honeymoon, the Hawkings began their new lives as Dr. Hawking, international physicist, and his dutiful wife, the college student.

NOTES

1. D. W. Sciama, *The Unity of the Universe* (Garden City, NJ: Doubleday and Company, 1961), p. vii.

2. William Bonnor, *The Mystery of the Expanding Universe* (New York: Macmillan Co., 1964), p. 163.

3. Alan Lightman and Roberta Brawer, *Origins: The Lives and Worlds of Modern Cosmologists* (Cambridge: Harvard University Press, 1990), p. 142.

4. Ibid., p. 141.

5. Jane Hawking, *Music to Move the Stars: A Life with Stephen Hawking* (London: Pan Books, 2000), p. 11.

6. Stephen Hawking, "Sixty Years in a Nutshell," in *The Future of Theoretical Physics and Cosmology*, ed. G. W. Gibbons, E. P. S. Shellard, and S. J. Rankin (Cambridge: Cambridge University Press, 2003), p. 106.

7. Kip S. Thorne, *Black Holes and Time Warps* (New York: W.W. Norton and Co., 1994), p. 272.

8. Lightman and Brawer, *Origins*, p. 144.

9. Jane Hawking, *Music to Move the Stars*, p. 18.

10. Muscular Dystrophy Association, *When a Loved One has ALS: A Caregiver's Guide* (Tucson: MDA, 2003), http://www.mdausa.org/publications/alscare (accessed September 10, 2003).

11. Morgan Strong, "Playboy Interview: Stephen Hawking," *Playboy* (April 1990): 68.

12. Stephen Hawking, *Black Holes and Baby Universes and Other Essays* (New York: Bantam Books, 1993), p. 166.

13. Stephen Hawking, ed., *Stephen Hawking's A Brief History of Time: A Reader's Companion* (New York: Bantam, 1992), p. 52.

14. Hawking, *Music to Move the Stars*, p. 37.

15. Ibid., p. 38.

16. Ibid., p. 44.

17. Ibid., p. 46.

18. Ibid., p. 72.

19. Ibid., p. 5.

20. Approximately five percent of ALS cases are hereditary.

21. Hawking, *Music to Move the Stars*, p. 72.

22. Strong, "Playboy Interview: Stephen Hawking": 68.

23. Tim Adams, "Brief History of a First Wife," *The Observer*, April 4, 2004, http://observer.guardian.co.uk/review/story/0,,1185067.html (accessed August 27, 2004).

24. Kitty Ferguson, *Stephen Hawking: Quest for a Theory of Everything* (New York: Franklin Watts, 1991), p. 45.

25. S. W. Hawking, "On the Hoyle-Narlikar Theory of Gravitation," *Proceedings of the Royal Society of London* A286 (1965): 313–19.

26. Lightman and Brawer, *Origins*, p. 419.

27. Stephen Hawking, "Sixty Years in a Nutshell," p. 111.

28. See, for example, S. W. Hawking, "The Occurrence of Singularities in Cosmology," *Proceedings of the Royal Society of London* A294 (1966): 511–21, and S. Hawking and G. F. R. Ellis, "Singularities in Homogeneous World Models," *Physics Letters* 17 (1965): 246–47.

29. Jane Hawking, *Music to Move the Stars*, p. 68.

30. Stephen Hawking, "Sixty Years in a Nutshell," p. 110.

CHAPTER 4

CHILDREN AND CALCULATIONS

Family Man and Theoretician

NEW BEGINNINGS, NEW CHALLENGES

After spending an all-too-brief and relatively inexpensive week-long honeymoon in Suffolk, the Hawkings took their first transatlantic flight. This was purely business—Stephen was to attend a summer school on general relativity at Cornell University in New York State. Their accommodations were in a dormitory where their hall-mates were all families with small, boisterous children. Stephen later admitted that the stressful situation definitely put a strain on their young marriage.[1] Nevertheless, Stephen made valuable contacts with leaders in the field and set the groundwork for his own emerging reputation.

It was at Cornell that Jane truly came face to face with the haunting reality of Stephen's disease, despite the fact that they never discussed it (at Stephen's request). After sharing conversation with friends in the chilly evening air, Stephen was seized with one of the choking fits common to his illness—the first Jane had ever witnessed. Shocked and helpless, Jane could do nothing until Stephen finally gestured for her to soundly thump him on the back. Jane later wrote that the "demonic nature of the illness had announced its presence much more dramatically than in lameness, difficulty of movement, and lack of coordination."[2] This was truly an alarming vision of the reality of the life that faced them.

October 1965 arrived, and with it Stephen's new position at Gonville and Caius College. With great difficulty (and no aid from the college), he and Jane had found a small house to rent near enough to Stephen's office for him to walk on his own. He would have to fend for himself during the week while Jane stayed at her own rented room at Westfield, continuing the final year of her college education. During the week Jane focused on her studies, and upon commuting home on Friday she would get right to the work of taking care of the house and typing up Stephen's thesis, only to leave again each Monday morning. Stephen remained acutely aware that his mathematical education was lacking (as compared to many of his colleagues), and devised a creative way to improve his own background while making much-needed additional money at the same time. He supervised an undergraduate math course for the college, teaching himself the techniques he lacked in the process.

Although the Hawkings did their best to lead normal lives, Stephen's increasing disability and England's lack of facilities for the handicapped seemed to conspire against them. A Sunday afternoon sightseeing trip to Anglesey Abbey turned into an exercise in frustration, as the staff would not allow them to park near the house despite Stephen's condition, demanding that they park in the official lot a half mile away. In response, Jane wrote a letter of protest to the director of the National Trust (who oversaw the abbey). This would be but the first in a long line of campaigns in which Jane and Stephen would be involved for the rights of the disabled.[3]

At the advice of Stephen's sister Mary, the Hawkings attended a December astrophysics conference in Miami and then spent a week visiting former fellow graduate student George Ellis and his wife in Austin, Texas. Stephen's choking spells were becoming more frequent and it was thought the warm, sunny weather would offer him some temporary relief. They returned in time for Christmas, and with their lease now expired, moved into another house on the same street (Little St. Mary's Lane) but slightly closer to Stephen's office.

That winter would bring them more good fortune, with the announcement that Stephen's essay "Singularities and the Geometry of Space-Time" was coawarded the Adams Prize (along with an entry by Roger Penrose). This prestigious mathematics prize offered by St. John's College of Cambridge University was named in honor of John Couch Adams, codiscoverer of the planet Neptune. It is given to a young researcher based in Britain, for work of international caliber. Dennis Sciama proudly told Jane that Stephen assuredly had a career worthy of Isaac Newton ahead of him.[4] A celebration for the official completion of Stephen's PhD was held in March 1966, and he became a member of the new Institute of Astronomy, housed outside Cambridge at the Observatory.

Jane completed her studies that spring, and although she did not get top honors on her exams, she did score highly enough to qualify for graduate school. She had seen enough of the social scene at Cambridge to know that many of the academics' wives had not been encouraged to fulfill their own potential and suffered emotionally because of it. She believed that in order to help Stephen, she had to first help herself by finding some outside intellectual identity and purpose. A PhD was the obvious step. She selected a topic that could easily be researched through library work alone (a necessity given Stephen's needs): a critical study of already published texts from medieval Spain. Understanding that she would not be able to complete the work within the three-year period demanded by Cambridge, she chose to remain affiliated with London University (of which her alma mater, Westfield, was a college).

FAMILY MATTERS

That decision turned out to be wise indeed, as Jane discovered she was pregnant that autumn, and Stephen's condition began to require greater care. At the advice of his father, Stephen began receiving vitamin B injections, administered by a local nurse who was willing to stop by once a week at breakfast time. Stephen's fingers were also beginning to curl, making handwriting all but impossible. Dennis Sciama was able to convince the Institute of Physics to fund twice-weekly physical therapy at the Hawking home in order to try and counteract the ravages of the disease on Stephen's joints and muscles.[5]

In March, both Jane and Mary Hawking officially received their diplomas from London University, and Mary assumed a medical practice in the eastern United States. Jane's thesis was largely put on hold as she prepared for a more immediate challenge—motherhood. Robert George Hawking came into the world on May 28, 1967, two weeks ahead of schedule. Less than two months later, Robert became the latest Hawking to cross the Atlantic, as he accompanied his parents to a seven-week summer school in Seattle, a two-week stay at the University of California at Berkeley, and a visit to Aunt Mary and Stephen's childhood friend John McClenahan.

The Hawkings returned to Cambridge after four months abroad to joyous news—Stephen's research fellowship had been renewed for two more years. Stephen's research was bearing significant fruit and his reputation was blossoming even further. Building on the mathematical techniques Penrose had developed to study singularities, Hawking and his colleagues began making predictions and issuing proofs relative to the existence and behavior of the big bang. For example, he and George Ellis proved that the observation of the cosmic background radiation verified the existence of the big bang.[6] He also worked with Robert Geroch, and Penrose himself, extending the singularity theorems to a variety of physical and mathematical cases. Hawking later recounted that it was a "glorious feeling having a whole field virtually to ourselves."[7] They began to analyze the global way that different points in the universe—different points in space-time—were causally related (the big picture as it were, rather than looking at the countless minute relationships between points), and found in the end that they could prove that the universe had to have an initial singularity, which was the very beginning of space and time. Hawking and Penrose wrote a joint essay on the beginning of time and took second place in the Gravity Research Foundation Award in 1968, and their ultimate paper on the subject, drawing together all previous singularity theorems, appeared in the *Proceedings of the Royal Society* in 1970.[8] Hawking summarized the paper as proving "that there must have been a Big Bang singularity provided only that general relativity is correct and that the universe contains as much matter as we observe."[9]

Although Hawking's creative future seemed secure, his prognosis for continued employment was far from rosy. His research fellowship had already been renewed once (considered the maximum), and in 1969 he was faced with the termination of his position. Given his physical disability and increasingly slurred and indistinct speech, he could not take on a teaching position where he would be expected to lecture. Old friends Dennis Sciama and Hermann Bondi came to his aid. A rumor surfaced that King's College was ready to give Hawking a Senior Research Fellowship, and Gonville and Caius stepped in with a special Fellowship for Distinction in Science with a six-year contract before the rival college could steal their rising star.

Elsewhere in the Hawking household, Jane was struggling to work on her thesis, thanks to brief respites of childcare provided by her mother and a neighbor's nanny. In early 1970 she found herself pregnant once again, and family priorities took precedence over research. They had managed to scrape together enough money for a mortgage deposit on their rented house, but it took time to find an institution willing to risk investing on a 200-year-old house that needed renovation. Renovations were finally completed a month before Lucy Hawking's birth on November 2. Caring for a toddler, a newborn, and an increasingly disabled husband proved a daunting task for Jane. Stephen could still pull himself up the stairs, but his walking had become so unsteady that he was finally relegated to a wheelchair. Dressing in the morning and undressing at night were slow and arduous processes, but ones that Hawking still insisted on managing mostly by himself. This left Jane some precious time for reading at night.

SHIFTING GEARS: BLACK HOLES

Before 1970, Stephen's main research interests had been the application of Penrose's work on the singularities of collapsing stars—what John Archibald Wheeler named "black holes" in a stroke of genius in the late 1960s—to the beginning of the universe and the big bang singularity. After completing his final paper with Penrose, Hawking turned his attention to the increasingly exciting field of black holes. Within a year he submitted three seminal papers on three different aspects of these gravitational monsters.

Hawking's first stroke of black hole genius was the proposal of an entirely new class of objects. Rather than being created in the violent collapse of dying stars several times the mass of our sun, these primordial black holes were merely the mass of a mountain and the size of a proton. They would have been squeezed into existence by the enormous pressures and energies of the early moments after the big bang.[10] Another avenue of his research was uniqueness theorems, the proof that, for all their fascinating gravitational strength, black holes are relatively simplistic beasts. Irregardless of the initial complexity of the material, which collapsed to form the black hole, after its creation it can be completely described by only three properties—its mass, overall electric charge (if any), and rotation. Theoretically, black holes created out of penguins, bowling balls, and marshmallows could appear identical—there would be no way of telling what the original material had been. In response, John Archibald Wheeler suggested, "black holes have no hair." It was this lack of distinguishing characteristics that Hawking helped to prove.[11] The third of Hawking's breakthroughs centered around a most unexpected connection between black holes and thermodynamics.

Thermodynamics[12] encompasses the laws of nature that regulate how large numbers of atoms (such as the air in a room) behave as a group in a statistical way. This includes the random motion caused by heat. Thermodynamics therefore includes the laws of heat and energy transfer between bodies of different temperatures. Demetrios Christodoulou, a young graduate student of John Archibald Wheeler, noticed the curious fact that some of the equations that governed black holes resembled equations from standard thermodynamics. In particular, Christodoulou noted that there was a certain property of the black hole that could never be decreased in any interactions the black hole might have—a mathematical combination of the black hole's mass and rotation (or, more properly, angular momentum) which he called the irreducible mass.[13]

According to the second law of thermodynamics, the entropy of a self-contained (closed) system always increases in any natural process. Entropy is related to the heat added to a system and its temperature, and is a statistical measure of the disorder of the system. The second law is therefore usually more simply stated: it is natural for a system to become more and more disordered. If this were not true then house repair contractors and cleaning services would be out of business. Another interpretation of entropy is that it represents the information that is unknown or missing about a system (such as the lack of a precise understanding of the state or arrangement of the atoms or molecules in an object). Was this apparent similarity between entropy and the irreducible mass a coincidence, or a clue of a deeper physical connection?

THE BOLD THINKER

Hawking had made a first step in answering this question around the time of Lucy's birth, suggesting to Penrose a new definition of the boundary of a black hole, known as the horizon. Hawking's definition was simply "the boundary between regions of spacetime that cannot send signals to the outside universe and those that can. Regions that can't communicate with the outside universe would be in the hole's *interior*, those that can communicate would be in the *exterior*."[14] This is now a commonly used definition. One night, after Lucy's birth, Hawking used the considerable time it took to get ready for bed to ponder the meaning of his new definition of the horizon (an absolute horizon). He realized that the horizon was made of light rays that were barely trapped by the gravitational field of the black hole, trying to escape but never actually getting any farther away, nor being drawn back in. He further understood that these captive light rays could not be approaching each other, because it would mean they would collide and fall into the black hole, thus meaning there would be a situation with no horizon and the internal singularity would be exposed to the universe—it would be naked. However, Roger Penrose had proposed the so-called cosmic censorship conjecture, which stated that every singularity is hidden (or clothed) inside a horizon. The only way to prevent the light rays from colliding with each other would be if the light rays were moving parallel to each other, or, even better, moving away from each other, but this would mean that the size (surface area) of the horizon would always have to increase, no matter what happened to the black hole. This is similar to the spines of a porcupine, which seem to spread out as it puffs out its body.

What would be the bottom line of this restriction on the size of the horizon? Hawking realized that when a black hole eats any material, or two black holes collide or otherwise interact in any way, the result must be that the total surface area of the horizons of all black holes involved must increase. Since the surface area is related to Christodoulou's irreducible mass, Hawking had found a more general and powerful statement of the graduate student's discovery. Realizing the importance of his discovery, Hawking was too excited to sleep well that night, and shared his discovery with Penrose first thing in the morning. The scientific community soon learned of this breakthrough in December at the Texas Symposium of Relativistic Astrophysics.[15] Hawking was careful to note that although the increase in surface area was certainly analogous to the increase of entropy, it was only an analogy and was not to be taken more seriously than that.

As he had in the past, Hawking wrote an essay, entitled "Black Holes," for the annual Gravity Research Foundation Award in January, but this time his work was selected for the top prize, and he received enough prize money to buy a newer car. His research continued to flourish, and he collaborated with James Hartle of the University of California, Santa Barbara, who was visiting the Institute of Astronomy. Together they derived a set of equations that explained how Hawking's horizons behaved and grew as the black hole consumed material.[16] As colleague Kip Thorne explains,

> Hawking is a bold thinker. He is far more willing than most physicists to take off in radical new directions, if those directions "smell" right. The absolute horizon smelled right to him, so despite its radical nature, he embraced it, and his embrace paid off.[17]

Despite the radical nature of his discoveries thus far, nothing could prepare him, or the scientific community, for his next bold stroke of brilliance.

NOTES

1. Stephen Hawking, *Black Holes and Baby Universes and Other Essays* (New York: Bantam Books, 1993), p. 17.
2. Jane Hawking, *Music to Move the Stars: A Life with Stephen Hawking* (London: Pan Books, 2000), p. 80.
3. Ibid., p. 88.
4. Ibid., p. 91.
5. Ibid., pp. 113–14.
6. S. W. Hawking and G. F. R. Ellis, "The Cosmic Black-body Radiation and the Existence of Singularities in Our Universe," *Astrophysical Journal* 152 (1968): 25–36.
7. Stephen Hawking, "Sixty Years in a Nutshell," in *The Future of Theoretical Physics and Cosmology*, ed. G. W. Gibbons, E. P. S. Shellard, and S. J. Rankin (Cambridge: Cambridge University Press, 2003), p. 111.
8. S. W. Hawking and R. Penrose, "The Singularities of Gravitational Collapse and Cosmology," *Proceedings of the Royal Society of London* A314 (1970): 529–48.
9. Stephen W. Hawking, *The Theory of Everything* (Beverly Hills: New Millennium Press, 2002), p. 42.
10. S. W. Hawking, "Gravitationally Collapsing Objects of Very Low Mass," *Monthly Notices of the Royal Astronomical Society* 152 (1971): 75–78.
11. S. W. Hawking, "Black Holes in General Relativity," *Communications in Mathematical Physics* 25 (1972): 152–66.
12. Appendix B reviews the four laws of thermodynamics.
13. Demetrios Christodoulou, "Reversible and Irreversible Transformations in Black Hole Physics," *Physical Review Letters* 25 (1970): 1596–97.
14. Kip Thorne, "Warping Spacetime," in *The Future of Theoretical Physics and Cosmology*, ed. G. W. Gibbons, E. P. S. Shellard, and S. J. Rankin (Cambridge: Cambridge University Press, 2003), p. 87.
15. S. W. Hawking, "Gravitational Radiation From Colliding Black Holes," *Physical Review Letters* 26 (1971): 1344–46.
16. S. W. Hawking and J. B. Hartle, "Energy and Angular Momentum Flow Into Black Holes," *Communications in Mathematical Physics* 27 (1972): 283–90.
17. Kip S. Thorne, *Black Holes and Time Warps* (New York: W.W. Norton and Co., 1994), p. 419.

CHAPTER 5

"STEPHEN'S CHANGED EVERYTHING"

Black Holes Aren't Black

FACING CHALLENGES

As Stephen's physical condition continued to slowly deteriorate, he and Jane continued to struggle against the barriers (both physical and emotional) they found in British culture. The passage of the Chronically Sick and Disabled Person's Act in 1970 did little to open doors for them immediately, as the government was slow to implement its own decree. Stephen and his family continued to find themselves battling the high curbs, steps, and endless staircases they encountered in family excursions as they attempted to live a rich and normal life. On one such outing, Stephen was dropped by two attendants at the Royal Opera House at Covent Garden who attempted to carry him up the stairs to their otherwise inaccessible seats. The Hawkings' activism for the rights of the disabled led to individual victories; for example, the Arts Theatre and Cinema eventually added seating areas for wheelchairs, and the English National Opera and the Coliseum also made accommodations. Even the university slowly began to make its campus more accessible.[1] Jane Hawking later noted that by "curious coincidence, the attitude of the City Council towards access for the disabled... mellowed rapidly as Stephen's fame grew."[2]

Hawking's entrance into black hole research had certainly made a splash. One of those who read the work with interest was Jacob Bekenstein, a graduate student at Princeton University working under John Archibald Wheeler. Bekenstein shared his advisor's discomfort at the bottom line of the no hair theorem for black holes. All detailed information about the forming material was lost (with the exception of mass, electric charge, and angular momentum). The same would, of course, be true about any material that later fell into the black hole. But what if someone tossed some entropy into the black hole? It would then appear that the entropy of the universe would decrease and the entropy of the black hole would increase—but there was no way to know for sure, because of the no hair theorem.[3]

Bekenstein decided that the solution was to take Hawking's thermodynamic analogy a step further, and as part of his thesis he made the radical (and immediately controversial) assertion that the surface area of a black hole *is* a measure of its entropy, thus allowing the second law of thermodynamics to be applied to black holes. There appeared to be a serious flaw to Bekenstein's reasoning—if a black hole had entropy, it had to have a temperature.[4] But any object with a temperature other than absolute zero must radiate. For example, the human body is certainly not hot enough to glow to the visible eye, but night vision goggles that detect infrared wavelengths clearly demonstrate that we radiate. The problem is that the very definition of a black hole is an object whose gravitational pull is so strong that light itself can't escape. How could such an object radiate? Clearly this was impossible (which Bekenstein himself admitted).

Hawking was openly irritated by Bekenstein's proposal, believing it "misused my discovery of the increase of the area of the event horizon."[5] He expressed his displeasure in a definitive way in August 1972 as a participant in Les Houches summer school on black holes in the French Alps, which Bekenstein also attended. Between lectures, Hawking teamed up with James Bardeen of Yale University, and his Institute of Astronomy officemate Brandon Carter, to write the definitive paper of the time on the way black holes worked—their mechanics. In the end they derived four laws of black hole mechanics, which surprisingly looked nearly identical to the classic four laws of thermodynamics "if one only replaced the phrase 'horizon area' by 'entropy,' and the phrase 'horizon surface gravity' by 'temperature.' (The surface gravity, roughly speaking, is the strength of gravity's pull as felt by somebody at rest just above the horizon.)"[6] The trio was careful to emphasize that these were only analogies, and in the final paper published the following year, they firmly stated that these four laws of black hole mechanics "are similar to, but distinct from, the four laws of thermodynamics."[7] Despite this pointed warning, Bekenstein was more convinced than ever that there was a true connection, although he could not make the leap of faith and claim that black holes had a true temperature, and therefore radiated. His advisor encouraged Bekenstein as best he could, saying "black hole thermodynamics is crazy, perhaps crazy enough to work."[8]

January 1973 also saw the completion of a project that Hawking had been working on with longtime friend and colleague George Ellis—a book on the details of the structure of space-time, "on length scales from 10^{-13} cm, the radius of an elementary particle, up to 1028 cm, the radius of the universe."[9] The book, which was based on Hawking's award-winning Adams Prize essay, was dedicated to the authors' former thesis advisor, Dennis Sciama, and soon became a classic text, although Hawking himself called it "highly technical, and quite unreadable."[10]

Around the Hawking household, new challenges were taking shape. Robert was now in school, and although he excelled in math, he took after his father in being a late bloomer in reading. Jane suspected her son was dyslexic, and knew he would receive a more specialized education better suited to his needs in a private school. However, Stephen's salary at Gonville and Caius, even combined with the research assistantships he subsequently received at the Institute of Astronomy and the DAMTP, was insufficient to cover such an expense. Fortunately, Stephen's father stepped in and offered to augment a family inheritance in order to buy a house, which Stephen and Jane rented out for profit.[11]

BLACK HOLES EXPLODE?

With the publication of his book, Hawking entered a new phase of his research career, moving from the study of classical general relativity to that of quantum gravity—the uneasy, tentative marriage between general relativity (the study of gravity and the structure of space-time) and quantum mechanics (the behavior of matter at the atomic or smaller scales) that has yet to be completely resolved. Indeed, a complete, seamless interweaving of general relativity and quantum mechanics (the two great revolutions of twentieth-century physics) is considered the Holy Grail of theoretical physics, as some call it a *theory of everything*. The big bang is an example of a phenomenon that, by definition, encompasses both theories—it deals with the universe as a whole (general relativity) but is also the source of all matter and energy in the universe (quantum mechanics). Another obvious uniting phenomenon was Hawking's primordial black holes—objects derived from general relativity, but because of their tiny size (that of a proton), that exist on the scale of quantum mechanics. Hawking reflected that his work with Roger Penrose "had shown that general relativity broke down at singularities. So the obvious next step would be to combine general relativity, the theory of the very large, with quantum theory, the theory of the very small."[12] Therefore, the application of quantum mechanics to black holes was an obvious next step in the evolution of Hawking's research agenda.

The groundwork for this new focus was laid in August and September of 1973. Stephen and Jane traveled to Warsaw, Poland, without the children, to attend a conference celebrating the 500th anniversary of the birth of the legendary Polish astronomer Nicolaus Copernicus. They then traveled to Moscow with Kip Thorne to meet with famed Russian theorist Yakov Zeldovich. Alexei Starobinsky, Zeldovich's graduate student, detailed research that he and his advisor had been conducting, which demonstrated that their own partial marriage of general relativity and quantum mechanics predicted that rotating black holes should radiate. The radiation was created by siphoning off the rotational energy of the black hole, which would cause the rotation to slow down until it stopped (and the radiation along with it). Hawking was intrigued by the idea but was not convinced by their proof, especially the method they had used to introduce quantum effects into general relativity. He resolved to explore the problem himself upon returning to Cambridge.

Although Hawking returned home with a fascinating problem with which to play, his physical limitations were taking an even greater toll on him and his family than ever before. His speech had become so slurred that some people simply found it impossible to understand him. Every physical activity required significant help, including eating and bathing. The stairs were becoming an increasing obstacle, and it was becoming harder and harder to pull himself up them (the activity itself a form of much-needed physical therapy).[13]

Turning his attention to his newfound topic of interest, Hawking realized that he "had no background in quantum theory... [s]o as a warm up exercise, I considered how particles and fields governed by quantum theory would behave near a black hole."[14] What his calculations predicted stunned Hawking—the young upstart, Bekenstein, had been on the right track after all. What Hawking discovered was that black holes *do* radiate (in the form of particles) which means that they do have a temperature, and that the four laws of black hole mechanics which he, Bardeen, and Carter had been so careful to stress as only analogous to thermodynamics, were actually laws of black hole thermodynamics in disguise. In hindsight, he explained, "I didn't want the particles coming out... I wasn't looking for them; I merely tripped over them. I was very sorry because it destroyed my framework, and I did my best to get rid of them. I was rather annoyed."[15] Initially, Hawking's calculations did not explain the mechanism for the radiation, merely the fact that it existed. Subsequently Hawking and others worked out a number of different but complementary derivations for this "Hawking radiation." Perhaps the most straightforward way to understand this involves a look at virtual particles.

According to Einstein's special theory of relativity, mass can be converted into energy (such as in an atomic bomb) and energy can be converted into mass (specifically a pair of massive objects, one being made of matter and the other its antimatter mirror image, or antiparticle). According to one of the most startling predictions of quantum mechanics (called the Heisenberg uncertainty principle), it is possible to create energy out of nothing so long as the energy debt is paid back in a very short period of time. This is similar to a bank loan, except that there is no interest, but the due date is absolute! This energy is, in a sense, borrowed from the very fabric of the universe—from the vacuum, as it is called. This happens randomly every second of every day, with the borrowed energy creating a particle-antiparticle pair (called a virtual pair, since they only exist temporarily). After the passage of a minute amount of time, the pair rejoins and annihilates, repaying the energy debt to the universe. As strange as this may sound, it has been verified experimentally. Parallel metal plates in a vacuum set very close together will feel a tiny attractive force due to the presence of the virtual pairs—this is called the Casimir effect.

Suppose the creation of a particle-antiparticle pair happened just outside the horizon of a black hole. If one of the members of the pair fell into the black hole, the remaining particle could not annihilate with its partner when the time was up, and the energy debt appears to be unpaid. A real particle or antiparticle has been created, seemingly from nothing, or, rather, it looks to an outside observer as if it leaked from the black hole! But the energy needed to create the particle had to come from somewhere—namely from the black hole itself. In this way, the energy of the black hole, or, better yet, its mass (since mass and energy are convertible, according to Einstein) decreases, and the black hole radiates away.

Hawking's equations told him that the rate of radiation (related to its effective temperature) was inversely proportional to the mass of the black hole—the more massive the black hole, the slower it should radiate. He calculated that for a black hole the mass of the sun, this should happen so slowly that the lifetime of such a black hole would be much longer than the age of the universe. Fortunately, he had earlier predicted the existence of tiny primordial black holes, with a mass much smaller. These objects could radiate away in a time span comparable to the present age of the universe. Since the rate of radiation increased as the black hole's mass got smaller and smaller, it would eventually lead to a runaway effect, with the black hole eventually exploding in a shower of gamma rays with the equivalent energy of roughly a million megaton thermonuclear weapons.[16] Gamma ray astronomers could, theoretically, observe this.

CRITICISM AND KUDOS

Hawking spent considerable time checking and rechecking his calculations, knowing full well how controversial and revolutionary his ideas were. "I worried about this all over Christmas, but I couldn't find any convincing way to get rid of them. In a way it was one of those accidental discoveries, like the discovery of penicillin."[17] In the end, he had to accept that what his equations were telling him was the truth. Einstein himself had felt uneasy about certain quantum mechanical predictions, especially their probabilistic nature—they predict the possibility of an occurrence but cannot determine precisely when and where something will happen. Einstein is quoted as saying "God does not play dice with the universe," to which colleague Neils Bohr countered that Einstein should not tell God what to do! Hawking now extended that admonition, finding that "God not only plays dice but also sometimes throws them where they cannot be seen"—namely into a black hole.[18]

Although careful to be discrete until he was ready to make an official announcement, he could not resist telling Dennis Sciama, the man who had helped shape his early career. Afterward, the shocking news spread through Hawking's inner circle. Martin Rees, a colleague at Cambridge, welcomed a visit from Sciama with "Have you heard? Have you heard what Stephen has discovered? Everything is different, everything has changed!"[19] In January Roger Penrose called Hawking while Hawking was eating his birthday goose dinner. Hawking later lamented that, although he appreciated Penrose's enthusiasm, he was rather disappointed that they talked so long that his dinner got cold.[20]

The official announcement was made in February 1974 at the Second Quantum Gravity Conference, held at the Rutherford-Appleton Laboratory at Oxford, and organized by Dennis Sciama. Upon the completion of the talk, the chairman of the session, John G. Taylor of Kings College, declared it to be nonsense. Soon after Hawking's paper on the discovery, "Black Hole Explosions?"[21] (the question mark reflecting Hawking's own lingering uncertainties) was published in the well-respected journal *Nature*, Taylor and P. C. W. Davies published an opposing paper in the same journal. They flatly stated that they believed that "it is a rather too literal interpretation of the concept of the 'temperature' of a black hole to apply it to emission processes in this way."[22] Yakov Zeldovich, whose own work on black hole radiation had sparked Hawking's interest in the subject, was also skeptical, but he was eventually convinced of the truth of the result.

Hawking wrote a longer, more complete description of his discovery and submitted it to *Communications in Mathematical Physics* in March 1974. He did not receive any acknowledgment, and a year later inquired as to the status of his paper. He was told that the journal had lost his manuscript and was asked to resubmit it. When it was finally published, the submission date was erroneously listed as April 1975.[23] Over the next two years, Hawking and others (including Jim Hartle in a joint paper with Hawking)[24] derived Hawking radiation in a number of different ways (as mentioned above), finally convincing all naysayers.

The discovery of black hole radiation sealed Hawking's international reputation as a top-notch researcher. Werner Israel noted that the discovery "crowned a classical theory which had reached an undreamed-of level of elegance and sophistication and opened the door to new horizons in the quantum domain."[25] John Archibald Wheeler proclaimed that talking about it was like "rolling candy on the tongue."[26] Hawking's calculations were made all the more remarkable by the fact that they were largely done in his head. Kip Thorne explains that

> because the loss of control over his hands was so gradual, Hawking has had plenty of time to adapt. He has gradually trained his mind to think in a manner different from the minds of other physicists: He thinks in new types of intuitive mental pictures and mental equations that, for him, have replaced paper-and-pen drawings and written equations.[27]

Hawking himself has a more modest explanation: that most people have "the mistaken impression that mathematics is just equations. In fact, equations are just the boring part of mathematics. I attempt to see things in terms of geometry."[28]

It came as little surprise when Hawking was selected for election to the prestigious Royal Society shortly after his announcement of black hole radiation. The spring of 1974 proved very auspicious indeed for the Hawking household. Besides the research triumph and the announcement of the major award to be bestowed upon Stephen, his son, Robert, was finally going to begin private school in the fall. In addition, an invitation arrived from Caltech (where friend Kip Thorne was a faculty member), offering him the Sherman Fairchild Distinguished Scholar visiting professorship for the 1974–1975 academic year. The position included not only a healthy salary, house, and car, but also an electric wheelchair and coverage of all other medical needs, as well as proper schooling for Lucy and Robert.[29] There was only one looming concern—Stephen's general care.

Jane single-handedly cared for Stephen as his condition continued to deteriorate, with the exception of the weekly vitamin shots and occasional physical therapy. As Stephen became more severely disabled, Jane found the responsibility of caring for her husband and children more difficult to manage on her own. Stephen later wrote that "no one was offering to help us, and we certainly couldn't afford to pay for help."[30] In addition, Jane understood that accepting outside help might force Stephen to admit the "gravity of his condition" and "his courage might fail him."[31] This was compounded by Hawking's intense focus on his work. Jane lamented that he

> could be very withdrawn and not talk for a whole weekend. . . . But his moods were such that I wouldn't know whether he was uncomfortable, worried about the illness or absorbed by physics. Then on Sunday evening his face would light up and it was "Eureka! I've solved a new equation."[32]

Over the Easter holiday, Jane devised a solution that would meet all their valid concerns—if one or more of Stephen's graduate students moved in with them (beginning with the trip to Caltech), a trusted friend could share the responsibility of providing Stephen with the care he needed. Thus it was that Bernard Carr agreed to accompany the Hawkings to Caltech that fall.

Hawking was inducted into the Royal Society on May 2, 1974. At the age of 32, he was one of the youngest inductees in its history. As the building was not handicapped-accessible, he was carried in and set into his wheelchair. It was tradition that newly elected members walk to the podium and sign their name in the official roll book, both tasks which defied his condition. The president, Sir Alan Hodgkin, a Nobel laureate in biology, brought the book down to Hawking to sign, which Hawking did with considerable difficulty. Standing in the doorway was Carl Sagan, famed American popularizer of astronomy. He was attending a meeting in another part of the building, and during a coffee break he had wandered into the ceremony out of curiosity. He later recounted watching Stephen laboriously sign his name, and the ovation that followed. As Sagan realized, "Stephen Hawking was a legend even then."[33]

NOTES

1. Jane Hawking, *Music to Move the Stars: A Life with Stephen Hawking* (London: Pan Books, 2000), pp. 177–78.

2. Ibid., p. 178.

3. Jacob D. Bekenstein, "Black Hole Thermodynamics," *Physics Today* (January 1980): 24–26.

4. From Chapter 4, entropy is dependent on the heat added to a system and its temperature.

5. Stephen W. Hawking, *The Theory of Everything* (Beverly Hills: New Millennium Press, 2002), p. 80.

6. Kip S. Thorne, *Black Holes and Time Warps* (New York: W. W. Norton and Co., 1994), p. 427.

7. J. M. Bardeen, B. Carter, and S. W. Hawking, "The Four Laws of Black Hole Mechanics," *Communications in Mathematical Physics* 31 (1973): 162.

8. Bekenstein, "Black Hole Thermodynamics": 28.

9. S. W. Hawking and G. F. R. Ellis, *The Large Scale Structure of Space-Time* (Cambridge: Cambridge University Press, 1973), p. xi.

10. Stephen W. Hawking, *A Brief History of Time* (Toronto: Bantam Books, 1988), p. vi.

11. Hawking, *Music to Move the Stars*, pp. 220–23.

12. Stephen Hawking, "Sixty Years in a Nutshell," in *The Future of Theoretical Physics and Cosmology*, ed. G. W. Gibbons, E. P. S. Shellard, and S. J. Rankin (Cambridge: Cambridge University Press, 2003), p. 113.

13. Hawking, *Music to Move the Stars*, p. 218.

14. Hawking, "Sixty Years in a Nutshell," p. 113.

15. Dennis Overbye, "The Wizard of Space and Time," *Omni* (February 1979): 46.

16. Stephen Hawking, "Black Hole Explosions?" *Nature,* 248 (1974): 30.

17. Overbye, "The Wizard of Space and Time": 106.

18. Stephen Hawking, "The Quantum Mechanics of Black Holes," *Scientific American* (January 1977): 40.

19. Stephen Hawking, ed., *Stephen Hawking's A Brief History of Time: A Reader's Companion* (New York: Bantam Books, 1992), p. 98.

20. Ibid., pp. 93–94.

21. Hawking, "Black Hole Explosions?": 30–31.

22. P. C. W. Davies and J. G. Taylor, "Do Black Holes Really Explode?" *Nature* 250 (1974): 37–38.

23. Stephen Hawking, *Hawking on the Big Bang and Black Holes* (Singapore: World Scientific, 1993), p. 3.

24. J. B. Hartle and S. W. Hawking, "Path-Integral Derivation of Black Hole Radiance," *Physical Review D* 13 (1976): 2188–2203.

25. Werner Israel, "Dark Stars: The Evolution of an Idea," in *300 Years of Gravitation,* ed. S. W. Hawking and W. Israel (Cambridge: Cambridge University Press, 1987), p. 265.

26. Bernard Carr, "Primordial Black Holes," in *The Future of Theoretical Physics and Cosmology,* ed. G. W. Gibbons, E. P. S. Shellard, and S. J. Rankin (Cambridge: Cambridge University Press, 2003), p. 236.

27. Thorne, *Black Holes and Time Warps,* p. 420.

28. Gerald Jonas, "A Brief History," *The New Yorker* (April 18, 1988): 31.

29. Hawking, *Music to Move the Stars,* p. 242.

30. Steven Hawking, *Black Holes and Baby Universes and Other Essays* (New York: Bantam Books, 1993), p. 168.

31. Hawking, *Music to Move the Stars,* p. 243.

32. Lisa Sewards, "A Brief History of Our Time Together," *Daily Telegraph,* April 27, 2002, http://www.lexisnexis.com (accessed August 19, 2004).

33. Carl Sagan, "Introduction," in *A Brief History of Time,* ed. Stephen Hawking (New York: Bantam, 1988), p. x.

CHAPTER 6

CALTECH AND CAMBRIDGE

Exploring New Horizons

PATH INTEGRALS AND PARADOXES

There was one significant detail to struggle with before the Hawkings and Bernard Carr could leave for Caltech. Although over the years Stephen had used the stairs in their house as a much-needed form of physical therapy, he had finally come to the point where he could no longer negotiate them. In addition, the house on Little St. Mary's Lane was simply too small for their growing family. Jane found a house already owned by the college, which could be made perfect for their needs. Previously used for undergraduate housing, the first floor alone was spacious enough for their family, and the floor plan could be easily navigated by wheelchair. However, the remaining floors could still be rented to students, who could have their own private entrance. Renovations and rent were negotiated, and Stephen and his entourage left for sunny California.

Better weather was not the only change Stephen's year at Caltech brought. Interacting closely with new colleagues, especially experts in particle physics, stimulated new areas of research and brought a fresh approach to problems he had already studied. Among the fertile minds he met at Caltech was Don Page, a graduate student. Together Hawking and Page wrote a paper proposing that primordial black holes might be observed exploding as gamma ray bursts.[1]

Hawking's groundbreaking work on radiating black holes was merely a first step in a much larger research agenda which was to drive Stephen for his entire career—the unification of gravity (Einstein's general theory of relativity) with the theory of the microscopic world (quantum mechanics). His goal—and that of all others involved in this quest—is "a complete understanding of the universe, why it is as it is and why it exists at all."[2] The same goal had previously eluded no less a genius than Einstein himself. The central problem remains the fact that the two theories—the very foundations of modern physics—are utterly incompatible in their original mathematical formulation and underlying philosophy. Michio Kaku of the City University of New York lamented that it is

> as if Nature had two minds, each working independently of the other in its own particular domain, operating in total isolation of the other. Why should Nature, at its deepest and most fundamental level, require two totally distinct frameworks, with two sets of mathematics, two sets of assumptions, and two sets of physical principles?[3]

Although a final solution to the problem was not on the immediate horizon in the mid-1970s, there existed mathematical approaches that gave approximations to quantum gravity in certain applications. The approach that Hawking personally found most promising was the sum over histories method developed by Nobel laureate Richard Feynman. This method used a mathematical technique known as path integrals, which seek to describe the probability of event A leading to event B by summing up all possible ways the system could have taken to get from A to B. Imagine a traveler trying to book a flight from New York to Los Angeles. One might expect that a direct flight between the two cities would be the most probable route. While this is certainly true, anyone who has actually taken the trip knows that it is also possible that there is a connecting flight, through any number of intermediate airports. There is a reasonable chance the traveler could have gone from New York to Los Angeles through Atlanta, Chicago, St. Louis, Denver, or Minneapolis, for example. It is far less likely that the weary traveler passed through London, Rome, or Hong Kong, but in these days of Internet travel bargains, one cannot completely rule it out. It is, however, out of the realm of possibility that the traveler reached Los Angeles via Mars or Jupiter, given the current status of space travel. Using this mathematical technique, Hawking and Jim Hartle of the University of California, Santa Barbara (who also spent time at Caltech that year) developed the "conceptually simple"[4] derivation of black hole radiation mentioned in Chapter 5.

It was at Caltech that Hawking first formulated what was to become another of many controversial ideas of his career—the black hole information paradox. Recall that black holes have no hair, in that after forming they are described by only three values—their mass, electric charge, and angular momentum. The other information about the material that formed the black hole could be said to be safely stored within the black hole, although inaccessible to an outside observer. However, what happened when a black hole radiated? Hawking showed that the black hole "emits with equal probability every configuration of particles compatible with conservation of energy, angular momentum, and charge." He called this "a quantum version of the 'no hair' theorems because it implies that an observer at infinity cannot predict the internal state of the black hole apart from its mass, angular momentum, and charge."[5] In other words, the black hole would not divulge the secrets of its origin as it died, and the information was completely and eternally lost to the universe, in direct contradiction to the predictions of quantum mechanics. Hawking flatly stated in his paper that this "is a great crisis for physics because it means that one cannot predict the future: One does not know what will come out of a singularity."[6] Although he submitted the paper containing these controversial results in August 1975, it was not accepted and published until November 1976. Hawking had initially given the paper the bold title "Breakdown of Physics in Gravitational Collapse," but it was published as the more conservative "Breakdown of Predictability in Gravitational Collapse."[7]

Once again, Hawking found himself at the center of a debate over deeply held fundamental beliefs concerning how black holes (and nature in general) behaved. Hawking insisted that his work suggested that quantum mechanics itself was essentially flawed; others argued that it was our incomplete understanding of nature (including a lack of a true unified theory) that made it appear as if there were a paradox. Hawking, however, was resolute in his conclusions. Nearly twenty years later, he boldly asserted in a public lecture that his colleagues would eventually come around to his line of thinking, "just as they were forced to agree that black holes radiate, which went against all their preconceptions."[8]

A FRIENDLY WAGER

Not all of his time in California was spent immersed in intricate calculations. In a bit of whimsy, Hawking and friend Kip Thorne engaged in a friendly wager that was to become the stuff of legend. Although theoreticians had been feverishly calculating the expected properties of black holes for a decade, observational evidence for the mere existence of these gravitational demons was slow in coming. In 1966, Russian physicists Zeldovich and Guseynov wrote a letter to the *Astrophysical Journal* suggesting that black holes might be found in spectroscopic binary star systems. In such systems, only one star is visible, and the existence of the other companion is inferred by the effect it has on the spectrum of (light emitted by) the visible star. In particular, as the two stars orbit each other, if their orbit is aligned in the right way, they will alternately seem to approach and recede as seen from Earth, leading to an alternating blueshift and redshift of the spectral lines because of the Doppler effect. The amount of shift, and its period, can be used to determine the combined mass of the two stars, from which the mass of the invisible companion can be roughly estimated. According to theories of stellar death, an invisible companion of more than two to two-and-one-half times the mass of the Sun is thought to be too heavy to be anything other than a black hole. The purpose of the Russians' letter was "to draw attention to the particular objects to which we have referred." They pointed out that candidate systems might be identified by the motion of gas circulating around the collapsed star by their "X-rays or other unusual spectral features."[9] Virginia Trimble and Kip Thorne published an article in *Nature* several years later analyzing specific candidate spectroscopic binaries,[10] with Hawking and Gary Gibbons making additional suggestions two years later.[11]

With the advent of x-ray astronomy in the 1960s, Zeldovich and Guseynov's suggestion could be used to further narrow down candidates. Such observations had to be conducted above the interference of the atmosphere, therefore requiring rockets or orbiting satellites. One of the first, and certainly the most infamous, objects found was Cygnus X-1. X-ray emissions from this area of the sky were first noted in 1962. The location of the source was more precisely pinpointed by the Uhuru x-ray satellite launched in 1971 and found to coincide with HDE 226868, a blue supergiant star. B. Louise Webster and Paul Murdin of the Royal Greenwich Observatory carefully monitored the spectroscopic binary and estimated that the mass of the unseen companion was between two and six times the mass of the Sun. Their published results ended with the statement that "it is inevitable that we should also speculate that it might be a black hole."[12]

Further observations refined the estimates of the star's mass, and by the mid-1970s there were some astronomers willing to bet that Cygnus X-1 was most definitely a black hole. Hawking and Thorne took this literally. The wording was carefully printed up, witnessed and signed by both men on December 10, 1974, and housed in Thorne's office for safekeeping. It read:

> Whereas Stephen Hawking has such a large investment in General Relativity and Black Holes and desires an insurance policy, and whereas Kip Thorne likes to live dangerously without an insurance policy,
>
> Therefore be it resolved that Stephen Hawking bets 1 year's subscription to "Penthouse" as against Kip Thorne's wager of a 4-year subscription to "Private Eye," that Cygnus X 1 does not contain a black hole of mass above the Chandrasekhar limit.[13]

Attesting to his wry sense of humor, Hawking bet against Cygnus X-1 in order to get the consolation prize of the magazine subscription, whereas if Cygnus X-1 did turn out to be a black hole, he would have the joy of knowing that his work on black holes had not been wasted. The 4-to-1 odds were based on the fact that "we were 80 percent certain that Cygnus was a black hole."[14]

Hawking also received several awards during his year at Caltech. In January 1975 he shared the Eddington Medal of the Royal Astronomical Society with friend and colleague Roger Penrose for "investigations of outstanding merit in theoretical astrophysics." Their groundbreaking work on singularity theories certainly embodied the award's guidelines. Several months later he flew to Rome with Bernard Carr to accept the Pius XI Gold Medal for Science from Pope Paul VI, the award reserved for a young scientist who had done distinguished work. The irony of the award was certainly not lost on Hawking or his friends. In 1633 Galileo was tried and condemned as "vehemently suspected of heresy" for violating a 1616 edict against work supporting the writings of Copernicus, which had in essence removed the Earth from the center of the universe. Now a scientist born exactly 300 years to the date from Galileo's death received accolades from the Vatican for work that also threatened to turn accepted ideas of the nature of the universe upside down. Jane wrote in a letter to friends that Stephen planned to use the occasion to "make a special plea for the rehabilitation of Galileo's memory."[15] Change was slow to come, but eventually Hawking received his wish. On November 10, 1979, at the Einstein Centenary meeting of the Pontifical Academy of Sciences, Pope John Paul II announced that the Vatican was reopening the case of Galileo.

Jane made a significant discovery of her own during their California sojourn. She accepted an invitation to join a chorus and found it "an intoxicating compulsion to my generally banal existence."[16] It would become both a source of strength and complications in the years to follow.

STRUGGLES AND SURPRISES

Complications of other sorts awaited the Hawkings' return to Cambridge. In Caltech Stephen had been able to use a fast, electric wheelchair, suitable for both indoor and outdoor transportation. He found it a tremendous asset to his independence and, as was within his rights, asked the Department of Health to provide one for his use in Cambridge. Although their new home was slightly farther from his office than his old house, it was still easily within commuting distance, but not in an old push-type wheelchair. His request was refused, and instead he was told he could apply for the same three-wheel electric car he had once used to travel between home and work. However, his physical condition had deteriorated to the point where he was not strong enough to operate that device, and an electric wheelchair was his only real option. Left with no choice, the Hawkings parted with their savings in order to purchase the much-needed equipment themselves.

Stephen's employment status at the university changed upon his return—for the better. The six-year contract for his special Fellowship for Distinction in Science was completed, and rumors had circulated that he was considering remaining in California permanently. In immediate response, the university offered him his first official post, a readership; he also received a secretary, the vivacious and dedicated Judy Fella, whom Jane described as introducing "a fresh vitality and an unaccustomed glamour to the drab realms of the Department of Applied Mathematics and Theoretical Physics." Judy remained an important and loyal fixture in Hawking's inner circle for many years.[17]

On the home front, more than just the address had changed. At a friend's insistence, Jane resumed work on her long-languishing thesis. She was also keen on continuing her singing and managed to squeeze in an hour per week for a voice lesson between caring for Stephen and the children and eking out work on her own research. Stephen's new postgraduate student, Alan Lapades from Princeton, moved into their spare room and added an extra pair of hands to Stephen's care. At the office, his students continued to provide much needed assistance, allowing him to finish up and publish the work he had done at Caltech.

The spring of 1976 proved disastrous, when Lucy, and subsequently Robert, came down with the chicken pox, and all three adults in the house developed severe sore throats and fevers. Stephen resisted seeing a doctor, understandably mistrustful of the medical profession given his experiences after his initial diagnosis with ALS. As a birthday present to Jane, he reluctantly allowed a home visit by their family doctor and was promptly whisked away to the hospital by ambulance. Although he returned home after a few days, he was plagued by nonstop, violent episodes of choking, remaining too weak to travel to work for several weeks. Jane became despondent, seeing no way their family could continue to care for Stephen's needs without significant outside aid. In her eyes, they could "no longer blind ourselves to the truth: we were living on the edge of a precipice, sometimes staring over the rim into the black void beneath."[18] Somehow she found the strength to continue, honoring Stephen's wish that his care be solely the domain of family, trusted friends, and students.

The year provided reasons to rejoice, as well. The Royal Society awarded Stephen the Hughes Medal "in recognition of an original discovery in the physical sciences" for his "distinguished contributions to the application of general relativity to astrophysics," and in the summer a BBC television crew filmed Stephen at home and while he gave a seminar in the department as part of a major documentary, *The Key to the Universe.* He also received the Dannie Heineman Prize for Mathematical Physics from the American Physical Society, which recognizes "outstanding publications in the field of mathematical physics."

Hawking and former student Gary Gibbons maintained a fruitful collaboration, directing their attention to the thermodynamic properties of de Sitter space. This solution to Einstein's field equations describes an empty universe that expands forever at a high rate of speed due to the presence of a "repulsive force."[19] The expansion is so fast that there are regions in this type of universe that an observer cannot see—ever! No matter how long the observer waits, light from these regions will never reach the observer's eye. These regions therefore lie beyond the observer's cosmological horizon. Gibbons and Hawking studied the quantum mechanical properties of this horizon and found that it was strikingly similar to the event horizon surrounding a black hole. In particular, it was endowed with an effective temperature and entropy and obeyed thermodynamic laws analogous to those for black holes. It also had its own kind of Hawking radiation, similar to a black hole. One can think of the de Sitter horizon like a black hole surrounding the observer, and as Gibbons and Hawking noted, the area of the horizon (which is related to the entropy) is "a measure of one's lack of knowledge about the rest of the universe beyond one's ken."[20] Their resulting paper was to become a seminal work in theoretical physics, demonstrating that the quantum properties of black holes which Hawking and others had discovered were applicable to the universe at large (at least in special geometric cases).[21]

During this time, Don Page was finishing his work at Caltech and exploring postdoctoral opportunities. Stephen managed to arrange a three-year North Atlantic Treaty Organization (NATO) fellowship at Cambridge for Don, and Don moved into the Hawking household shortly before Christmas. Don and Stephen were compatible in their physics, but their worldviews were like the opposite poles of a magnet. While Hawking was an unabashed atheist, Page was an evangelical Christian and tried to literally proselytize at the breakfast table. While Page soon realized it was a losing battle, it was not to be the only time that religion played a divisive role in Hawking's inner circle.

Having Don as Stephen's de facto assistant reaped unexpected benefits. Jane was able to remain at home rather than having to travel with Stephen. During the summer of 1977, Stephen and Don returned to Caltech for several weeks. While there, Stephen continued to work on quantum gravity. Since the path integral approach had proven useful in studying black hole radiation, he believed it would yield further fruit in developing a more all-encompassing marriage between quantum mechanics and general relativity. The result was a paper in which he acknowledged that although there was not yet a "satisfactory quantum theory of gravity" (what he called the "most important unsolved problem in theoretical physics today"), he firmly believed the path integral method was "the most suitable for the quantization of gravity."[22]

Hawking returned home to a promotion, a special chair in gravitational physics. He was now considered a professor and was given a raise in salary in recognition of the achievement. Jane was invited to join the local St. Mark's Church choir for the holiday concert season, which was directed by Jonathan Hellyer Jones. Jonathan, who was several years younger than Jane, had lost his wife of one year to leukemia eighteen months earlier. He and Jane became friends, and he began to visit the Hawking home, teaching Lucy to play the piano and volunteering to help with Stephen's physical needs. Jane had found a much-needed confidant and friend, and another freely offered set of hands to alleviate some of her burden. She officially joined St. Mark's parish and found additional strength in the sage and patient advice of their vicar, Bill Loveless. With this renewed sense of optimism, she began work on the final chapter of her thesis.

Honors continued to be heaped upon Stephen. He was given the Albert Einstein Award by the Lewis and Rosa Strauss Memorial Fund in Washington, D.C. This infrequently conferred honor is the highest American award given to physicists. He also received an honorary doctorate from the University of Oxford in the summer of 1978. With these awards came renewed interest from the scientific press. Journalists seemed as fascinated by his physical limitations as his physics, and the resulting articles seemed to put these two facets into curious juxtaposition. *New Scientist* exclaimed that it was a "poetic irony that Hawking can mentally roam the immensities of space and time in liberation from his physical limitations,"[23] while *Time* marveled at how he could do calculations without writing them down, "a feat that his colleague Werner Israel says is equivalent to Mozart's having composed an entire symphony in his head."[24] A lengthy article in *Omni* boldly declared him to be "The Wizard of Space and Time." Hawking explains to the author, noted science writer Dennis Overbye, that when it came to black holes, "I feel in a sense that I'm their master."[25]

Overbye's article introduced a wider audience to Hawking's humor as well as his science. He explained that he had once submitted a paper to the august *Physical Review* containing the line "Suppose you have a little race of gnomes." Their editors changed it to "Suppose you have observers."[26] Other examples remained secreted within Hawking's circle at the DAMTP. In one instance, Hawking had begun using the acronym ALE for Asymptotically Locally Euclidean, a nod to his students' ongoing battle to get the local pubs to serve what they considered real beer.[27]

Between other projects, Hawking and Werner Israel labored on a special undertaking, a volume of solicited articles on current research areas in general relativity, in honor of the centenary of Einstein's birth coming the following year. In their introduction to the volume, they acknowledged the ultimate aim of the featured research was to realize "Einstein's dream of a complete and consistent theory that would unify all the laws of physics."[28]

In September, Jane discovered she was pregnant for the third time and was determined to finish her thesis before the baby's arrival the following spring. Having Jonathan to help with household chores and Stephen's needs made this at least a remote possibility. During that winter, Stephen became a patron of the newly formed Motor Neuron Disease Association, and Jane and Jonathan organized a charity recital, which would be repeated several times over the following years. Timothy Hawking arrived on schedule, in April 1979, on Easter Sunday.

This was not the only cause for celebration in the Hawking home that year. Stephen was appointed to the famous Lucasian Chair in Mathematics at Cambridge, effective in November. The Reverend Henry Lucas, the university's representative to Parliament, had endowed the chair in 1663. The prestigious chair had been once held by none other than Sir Isaac Newton. Hawking would be the seventeenth holder of the chair, which came with a curious and rather appropriate edict—that the holder not be active in the church. Once more Hawking and religion had crossed paths and just as quickly they diverged. This tradition was soon to come to an unexpected end.

NOTES

1. D. N. Page and S. W. Hawking, "Gamma Rays from Primordial Black Holes," *Astrophysical Journal* 206 (1976): 1.

2. John Boslough, *Stephen Hawking's Universe* (New York: Quill/William Morrow, 1985), p. 93.

3. Michio Kaku, *Introduction to Superstrings and M-Theory*, 2nd ed. (New York: Springer-Verlag, 1999), p. 3.

4. J. B. Hartle and S. W. Hawking, "Path Integral Derivation of Black-hole Radiance," *Physical Review D* 13 (1976): 2188.

5. S. W. Hawking, "Breakdown of Predictability in Gravitational Collapse," *Physical Review D* 14 (1976): 2462.

6. Ibid.: 2460.

7. Faye Flam, "Plugging a Cosmic Information Leak," *Science* 259 (1993): 1824.

8. Stephen Hawking and Roger Penrose, *The Nature of Space and Time* (Princeton: Princeton University Press, 1996), p. 37.

9. Ya.B. Zeldovich and O. H. Guseynov, "Collapsed Stars in Binaries," *Astrophysical Journal* 144 (1966): 840.

10. V. L. Trimble and K. S. Thorne, "Spectroscopic Binaries and Collapsed Stars," *Astrophysical Journal* 156 (1969): 1013–19.

11. G. W. Gibbons and S. W. Hawking, "Evidence for Black Holes in Binary Star Systems," *Nature* 232 (1971): 465–66.

12. B. Louise Webster and Paul Murdin, "Cygnus X-1—a Spectroscopic Binary With a Heavy Companion?" *Nature* 235 (1972): 38.

13. Kip S. Thorne, *Black Holes and Time Warps* (New York: W. W. Norton and Co., 1994), p. 315.

14. Stephen W. Hawking, *The Theory of Everything* (Beverly Hills: New Millennium Press, 2002), p. 66.

15. Jane Hawking, *Music to Move the Stars: A Life with Stephen Hawking* (London: Pan Books, 2000), p. 263.

16. Ibid., p. 267.

17. Ibid., p. 281.

18. Ibid., p. 304.

19. The basic properties of de Sitter space are discussed in Appendix A.

20. G. W. Gibbons and S. W. Hawking, "Cosmological Event Horizons, Thermodynamics, and Particle Creation," *Physical Review D* 15 (1977): 2739.

21. This is true in space-times with a positive vacuum energy/cosmological constant. Consult Appendix A for information on the cosmological constant.

22. S. W. Hawking, "Quantum Gravity and Path Integrals," *Physical Review D* 18 (1978): 1747.

23. Ian Ridpath, "Black Hole Explorer," *New Scientist* (May 1978): 309.

24. "Soaring Across Space and Time," *Time* (September 4, 1978): 56.

25. Dennis Overbye, "The Wizard of Space and Time," *Omni* (February 1979): 107.

26. Ibid.: 104.

27. Chris Pope, "57 Varieties in a NUT Shell," in *The Future of Theoretical Physics and Cosmology*, ed. G. W. Gibbons, E. P. S. Shellard, and S. J. Rankin (Cambridge: Cambridge University Press, 2003), p. 517.

28. S. W. Hawking and W. Israel, ed., *General Relativity* (Cambridge: Cambridge University Press, 1979), p. xvi.

CHAPTER 7

PHYSICS OR METAPHYSICS?

The "No-Boundary" Proposal

LAUDABLE GOALS

Nineteen eighty arrived on an ominous note, as a stubborn cold for Stephen turned more serious. Jane was under the weather herself, and the stress of caring for her baby and her ill husband no doubt hindered her body's power to heal. Their family doctor recommended that Stephen spend some time in a local nursing home, to allow both him and his family time to rest and gather their strength. While Stephen was recuperating, Martin Rees, the Plumian Professor of Astronomy, requested a meeting with Jane at the Institute of Physics. Rees had succeeded Fred Hoyle in the position in 1973 and, having arrived at Cambridge two years after Hawking began graduate school, was painfully aware of Stephen's deteriorating condition and of the strains, both personal and financial, that the future would bring. He therefore offered to find funding to provide limited home nursing care for Stephen.

Although Stephen was initially displeased with the increasing intrusion into his life,[1] he soon realized that the presence of trained nurses could actually liberate him. Not only would they attend to his needs at the beginning and end of each day, but they could eventually be called upon to travel with him to conferences, thus affording him some manner of independence from his family, friends, and students. The burden placed upon his inner circle would be somewhat relieved as well.

The Hawking children were acutely aware that their family was different from those of their friends, but Jane struggled to provide as normal an environment as possible. Lucy recalled that she had a "normal, happy childhood, thanks to the heroic efforts of her mother and the sweetness of her maternal grandparents."[2] However, her father's condition was an undeniable constant in her life. He had become relegated to a wheelchair by the time she was born and was unable to play with his children in the physical rough-and-tumble way most families enjoy. It was considered a special treat when he would "be persuaded to wiggle one of his ears."[3] After her mother had tried to explain ALS to her, Lucy cried, thinking "he was going to die the next day." She always had the worry that "he would die soon. We lived life in the present tense." Lucy and her siblings also had to deal with the rude stares of strangers, to which she would respond by staring back "to see how they liked it."[4] She also

The Hawking family, 1980. From left to right, Timothy, Robert, Lucy, Stephen, and Jane. Time Life Pictures/Getty Images.

found herself sometimes acting as her father's interpreter in the presence of strangers, as his speech became more and more soft and slurred. Despite the pressures on her, Lucy had only one sincere wish—to "make my dad walk. I desperately wanted to change reality and make everything better."[5]

On April 29, Stephen was formally inaugurated as Lucasian Professor of Mathematics, and in keeping with a somewhat archaic tradition, presented an inaugural speech (read by one of his students). The title, "Is the End in Sight for Theoretical Physics?," was provocative, its basic premise dangerously boastful. He boldly stated that he expected the achievement of a complete unification between quantum mechanics and general relativity to be completed by the end of the twentieth century. On an alarmist note, he ended his speech by predicting that computers would overtake the human mind in terms of the ability to do theoretical physics, and that "maybe the end is in sight for theoretical physicists, if not for theoretical physics."[6] There was one unfinished piece of traditional business connected with his new chair. There had apparently been an oversight concerning a "big book, which every university teaching officer is supposed to sign." It was realized more than a year after he had taken the office that he had never signed the book. The volume was brought to his office, and Stephen recalls, "I signed with great difficulty. That was the last time I signed my name."[7]

While Stephen had achieved what appeared to be the height of his career, Jane was also attaining laudable goals. She passed an oral exam for her thesis in June 1980 and officially graduated in April 1981. After receiving her degree, Jane was asked to tutor local children in French and eventually received a part-time position at the Cambridge Centre for Sixth Form Studies, preparing students for their university entrance exams.[8] She had at last found a rewarding outside activity that utilized her intellectual skills and allowed her to work out of her home as well as spend time with her children and husband.

The Hawkings traveled to Rome in late September, where Stephen was presenting at the Study Week on Cosmology and Fundamental Physics hosted by the Pontifical Academy of Science. The academy had begun hosting these conferences in 1948 as part of its duties, and this was the third dedicated to astronomical topics. This particular session was seen, in part, as a tribute to a former president of the academy, Georges Lemaître, one of the fathers of the big bang model.[9] Stephen's nurses were not yet traveling with him, so his new postdoctoral researcher, Bernard Whiting from Australia, accompanied the Hawkings to the conference.

Stephen's paper was based on work he had been pondering for some time, the application of Feynman's sum over histories method to the universe at large. This involved summing all the possible pasts, or histories, the universe could have had, which was equivalent to summing over all possible space-times. One of the quandaries that faced someone doing such a calculation was to select the appropriate boundary conditions for the mathematical calculations. Boundary conditions refer to the beginning or initial state of a physical system, or what special circumstances must be considered in order to perform the calculations. His Vatican paper was therefore entitled "The Boundary Conditions of the Universe," and in it he made the bold assertion that as a boundary condition, only models of the universe that curve back in on themselves—so-called compact metrics—should be included in the calculations. In such a way, there would be no edge that needed to be specified and no unusual initial starting conditions to cause difficulties in the calculations. He argued that there "ought to be something very special about the boundary conditions of the universe and what can be more special than the condition that there is no boundary."[10]

At the end of the conference, the attendees had a special audience with Pope John Paul II. He addressed the scientists, admonishing them that the answer to the beginning of the universe could not be resolved by pure science, but instead that knowledge "comes from the revelation of God."[11] Hawking later recounted that "I was glad then that he did not know that the subject of the talk I had just given at the conference was the possibility that space-time was finite but had no boundary, which means that it had no beginning, no moment of creation."[12] Hawking realized he had come upon a possibly powerful technique, but since he did not know how to progress further with it to make precise predictions about the universe, he set it aside and focused on a new and equally exciting theory—inflationary cosmology.

INFLATION

In 1980, most scientists believed the correct description of the universe rested upon two theoretical legs—the big bang (derived from general relativity) and the so-called standard model of particle physics (derived from quantum mechanics). Both had withstood numerous observational tests, but both theories had nagging difficulties as well.[13] In 1980, Alan Guth, a young particle physicist at Stanford University, published a groundbreaking paper in which he suggested a modification to the big bang model that might solve some of those lingering problems. It became known as inflationary cosmology (or inflation for short) because it predicted that the infant universe, merely a fraction of a second old, had undergone a brief period of tremendous growth, inflating from smaller than an atom to the size of a baseball or larger in the blink of an eye.

Inflation is based on the concept of phase transitions, similar to water freezing to form ice. In the indescribably hot conditions that existed right after the big bang, the four fundamental forces of nature—gravity, electromagnetism, the strong nuclear force, and the weak nuclear force[14]—would be united in one superforce. It was this theoretical state of unity that quantum gravity (or any other theory of everything) hoped to describe. As the universe cooled, one by one the forces would become separate and distinct from each other, or would freeze out. Gravity, the most distinctive of the forces, would freeze out first, at approximately 10^{-43} seconds after the big bang. Next, the strong nuclear force would follow suit, at approximately 10^{-35} seconds after creation. Just as an ice cube does not freeze all at once, pieces of the universe would go through the transition bit by bit, in the form of bubbles. However, what if these bubbles did not immediately undergo this change of state at the critical temperature? What if they "forgot to freeze"? Such a phenomenon has been observed in water—supercooling. An example is freezing rain; liquid drops fall through the atmosphere during winter storms even though the air temperature is below freezing. However, this is a temporary state of forgetfulness, and the water must eventually make the transition to the frozen state. In the case of freezing rain, this occurs when the water strikes power lines, a tree branch, or the sidewalk, instantly creating a dangerous sheet of ice. Likewise, the bubbles in the early universe must complete their transition as well, and the strong force must separate. However, while the bubble is in this strange supercooled state, it has the curious property that it inflates at a phenomenal speed under the infiuence of a repulsive force, much faster than the rate of expansion observed today in Hubble's law; hence the name of the theory—inflation.

Guth proposed that after the bubbles underwent their tardy phase transition they collided, coalesced, and reformed into the normal structure of the universe, now expanding at a far more leisurely (and predicted) rate. However, Guth himself admitted that getting the bubbles to do this in the correct way was a weak spot in his theory. Regardless of this limitation, the theory took the physics world by storm, and countless particle physicists and cosmologists found themselves working on the details to Guth's brilliant, if incomplete, proposal. There was one especially intriguing feature of his model, the prediction that it would generate density perturbations. In layperson's terms, these are areas of slightly higher than average density where material is slightly clumped together. These small imperfections in the otherwise smooth distribution of matter and energy in the early universe are of vital importance, as they were the seeds for the eventual formation of galaxies and other structures. If inflation could provide the mechanism for generating seeds of the correct size and distribution, one of the most important unanswered questions about the big bang could be solved.

Hawking had first worked on density perturbations in an expanding universe in 1966.[15] It was therefore not surprising that Hawking was among those who sought the answer to that question. However, Hawking and his collaborators discovered that inflation (at least in the form proposed by Guth) would generate perturbations much too large to give rise to the universe we see today. In their acknowledgments, they thanked Hawking's 14-year-old son Robert for the "first version of the computer program used for the numerical calculation."[16]

In October 1981, shortly before this paper was submitted for publication, Hawking traveled to a quantum gravity conference in Moscow to deliver a paper on his findings. Among those in attendance was Andrei Linde, a Soviet physicist. Linde had independently come upon the same basic idea as Guth, but because of the rigid rules of censorship governing scientific publications in the Soviet Union, he was not able to publish his ideas until well after Guth. Linde understood the limitations of the original inflationary model and had devised a new inflation,[17] which he believed solved the problems. In this new model, the bubbles did not reconnect after inflating, but instead evolved independently of each other, our entire visible universe deriving from just one of these pockets of the entire big bang. Linde unveiled his improved model at the Moscow conference, and it was instantly met with considerable interest.

While translating the question and answer session after Hawking's talk, Linde found himself in the most awkward of situations, as Hawking proceeded to tear apart his theory. After it was over, Linde and Hawking found a deserted office in which to discuss it privately, "and for almost two hours the authorities of the Institute were in a panic searching for the famous British scientist who had miraculously disappeared."[18] The debate was later continued in Hawking's hotel room, a rather curious beginning to what would become an enduring friendship.

Hawking's next travels were to Philadelphia, where he received the Benjamin Franklin Medal in Physics for outstanding achievement from the Franklin Institute. Previous recipients of this award included Albert Einstein and Edwin Hubble. In his address to the institute he warned of the proliferation of nuclear weapons in the United States and USSR, and of the dangers escalating nuclear stockpiles posed to life on Earth. This was a topic he and Jane shared as a common point of concern, going so far as joining Newnham Against the Bomb, their local chapter of the Campaign for Nuclear Disarmament.[19] This would not be the last time Hawking would speak out publicly on this vitally important issue.

Upon returning to Cambridge, Hawking discussed Linde's ideas with graduate student Ian Moss. To his surprise, Hawking was sent Linde's paper[20] to review for publication. He was honest about its flaws but stated that the basic idea was certainly worthy of publication, especially as it would take far too long for Linde to make the necessary revisions within the confines of the oppressive Soviet system.[21] Instead, he and Moss submitted their own brief paper outlining a possible solution to the problem of how to properly end the inflationary period.[22] Meanwhile, Hawking and Gary Gibbons sent out invitations to several dozen colleagues inviting them to a special workshop on theories of the first minute of the universe (especially inflation) to be held the following summer.

The new year found Hawking on the prestigious Honours List, slated to become a Commander of the British Empire. The investiture was held at Buckingham Palace on February 23, with son Robert acting as Stephen's assistant when Queen Elizabeth presented the award. During the year he received honorary degrees from Notre Dame, the University of Chicago, and Princeton, among others.

The Nuffield Workshop on the Very Early Universe was held from June 21 to July 9, 1982, bringing together some of the greatest minds in inflationary cosmology, including Alan Guth, Andrei Linde, Paul Steinhardt, and Michael Turner. Its goal was to iron out some of the details of new inflation, especially those related to the still-bothersome density perturbations. For his part, Hawking was able to demonstrate that the inherent temperature of de Sitter space, which he and Gary Gibbons had predicted, would lead to small density perturbations.[23] Alan Guth later described the meeting as "among the most exciting times of my life. I will always cherish the memories of those events, and I will always be grateful to Stephen for the important role that he played in making them happen."[24]

The success of the conference was due in part to the remarkable dedication of Judy Fella. Not only was she Stephen's secretary, but she acted in the same role for the entire relativity group. In the days before flextime was the vogue, Stephen kept a rather unorthodox schedule, arriving late in the morning and remaining at work until seven or eight at night. Judy adjusted her own schedule to match Stephen's needs. She was the one responsible for the administrative details of the Nuffield Workshop as well as the later published proceedings edited by Hawking and colleagues.[25] Her dedication was, understandably, greatly appreciated by Stephen.[26]

IMAGINARY TIME

After the Nuffield Workshop, Stephen flew to California to spend the remainder of the summer at the newly created Institute of Theoretical Physics at the University of California, Santa Barbara. There he discussed his no-boundary proposal with longtime friend Jim Hartle. Hartle agreed that the choice of compact metrics was the correct one for describing the early universe within the context of the sum over histories technique. Together they worked out how to describe the quantum state of the universe using the no-boundary proposal. They summed over all possible states of a specific kind, "curved spaces without singularities, which were of finite size but which did not have boundaries or edges. They would be like the surface of the earth but with two more dimensions."[27] This choice of spaces "without singularities" seemed to directly contradict the earlier work of Hawking proving that the universe had an initial singularity. However, this apparent paradox was due to a rather subtle, yet important, aspect of the calculations.

Path integrals of the type Hartle and Hawking were trying to calculate are much more manageable if they are done by measuring time in imaginary numbers—what is called imaginary time. Hawking admitted that this is a "difficult concept to grasp" and later wrote that this was the main point of difficulty for readers of his popular-level works.[28] Imaginary numbers are best described by considering the mathematical function of the square root. If one multiplies a number by itself, one gets the number's square. For example, $2 \times 2 = 4$, so 4 is 2 squared. Looking at the process in reverse, it means that 2 is the square root of 4. However, $(-2) \times (-2) = 4$ as well, so -2 is also the square root of 4. The question now becomes, what is the square root of -4? In other words, what number, multiplied by itself, gives –4? The answer is that there is no real number with that property. Therefore, mathematicians invoke imaginary numbers in order to handle these calculations. The basic imaginary number is *i* and is defined as the square root of –1. In this way, the square root of –4 is 2*i*, the square root of –9 is 3*i*, and so forth.

In the equations of general relativity, time and space are connected, yet not treated exactly the same. By using imaginary numbers to describe time, space and time are mathematically identical. This is called a Euclidean space-time and the technique is referred to as the Euclidean approach. The Euclidean space-times chosen by Hartle and Hawking were closed surfaces, finite and unbounded (in imaginary time)—hence the no-boundary description. Hawking realized that there were

> fundamental implications for philosophy and our picture of where we came from. The universe would be entirely self-contained; it wouldn't need anything outside to wind up the clockwork and set it going. Instead, everything in the universe would be determined by the laws of science and by rolls of the dice within the universe.[29]

In imaginary time, this finite universe would have a definable beginning and end, like the North and South Poles of the Earth. Like the Poles, although we can define them, they are normal points in space, and the laws of physics would hold, just as the principles of geography still hold at the Poles. However, asking what happened before the big bang would be as ridiculous as asking what is farther north than the North Pole. Hawking suggested that "instead of talking about the universe being created, and maybe coming to an end, one should just say: The universe is."[30] In real time, though, the singularity would still exist. It would appear on the surface as if this mathematical trick is of no use; however, if it allows physicists to make calculations that they would not otherwise be able to do, then it has proven quite useful. In the end, the imaginary history and the real history are connected to each other through the language of mathematics, and the calculation of one should allow for the discovery of the other.

Hartle and Hawking were working with an admittedly simplified version of the universe, called a "minisuperspace," which could not reproduce all the properties of the real universe. They closed their paper with the suggestion that a more realistic (and complicated) mathematical model might actually provide a model of "the quantum state of the observed Universe," as well as the Zen-like pronouncement that "one would have solved the problem of the initial boundary conditions of the Universe: the boundary conditions are that it has no boundary."[31] The initial paper and subsequent work drew criticism from philosophers, who challenged the use of imaginary time. Hawking was firm in his response: "I think these philosophers have not learned the lessons of history.... I want to suggest that the idea of imaginary time is something that we will also have to come to accept."[32] He has made even bolder assertions, suggesting that perhaps this strange "Euclidean space-time is the fundamental concept and what we think of as real space-time is just a figment of our imagination."[33]

Regardless of its controversial nature, Hartle and Hawking's paper became one of the most widely cited papers in the field, ranking second on a list of citations on the Stanford Linear Accelerator Center's online archive (behind Hawking's follow-up to his initial black hole radiation paper). Of the eighty-eight papers listed on the all-time list, eleven were written or cowritten by Hawking.[34] As with any scientific theory, the no-boundary proposal made predictions that can be experimentally tested, namely a specific pattern of density perturbations reflected in the slight fluctuations in temperature in the cosmic background radiation. These predictions await definitive testing from satellites scheduled to be launched in the coming years. However, one primary prediction of the no-boundary proposal appeared to be in open contradiction with observation—it predicted that the universe would one day collapse in on itself, leading to the big crunch, or a closed universe.

SUPERSYMMETRY

To his credit, Hawking did not put all his theoretical eggs in one basket. He was open to any techniques and models that might lead to a better understanding of the universe. One of the theories he found especially promising was supergravity. This theory was an outgrowth of studies of supersymmetry, a unification theory that tried to match up all fundamental particles in nature with supersymmetric partners—particles with the same mass but different intrinsic spin.

Physicists picture subatomic particles like little tops, with the special property that their spin is "quantized"—it can only take on particular values, which are multiples of a fundamental unit of spin. The electron, for example, has exactly this particular value of spin, called spin 1/2. This gives the electron the very peculiar property that it has to complete two rotations in order to return to its original configuration. Other particles could have spin 3/2, most notably the gravitino, the supersymmetric partner of the graviton (the hypothetical particle associated with the gravitational force), or even spin 5/2. Particles with such half-integral values of spin are called fermions and have the quantum mechanical property that no two identical fermions can occupy the same quantum state. This is called the Pauli exclusion principle and is analogous to the social convention that students in a classroom all sit in different chairs. The Pauli exclusion principle is the source of the rule taught in general chemistry classes that electrons fill up the quantum states of an atom one at a time, beginning with those closest to the atomic nucleus.

There also exist particles with integral spin, for example, values of spin equal to whole numbers such as 1 and 2. A particle of spin 1 (like a photon, a particle of light) would return to its original configuration after completing one rotation (unlike the electron). A particle of spin 0 is called a scalar particle and must have a mass (unlike the photon, which is massless). Particles with integral values of spin are called bosons and do not obey the Pauli exclusion principle. This means that as many bosons can pack into a single quantum state as they wish. In the classroom analogy used previously, as many students would be allowed to sit in the same seat as they wished. This means that photons can "pig-pile" on each other to their hearts' content, resulting in more and more intense beams of light.

Nature therefore appears to have two different types of particles, bosons and fermions. What is the difference in their purpose? Fermions are particles that make up matter (such as electrons, protons, and neutrons[35]), while bosons include the particles that seem to carry, or mediate, the four fundamental forces of nature. For example, the electromagnetic force is mediated by the photon. Supersymmetry provides every fermion with a supersymmetric partner that is a boson, and vice versa, thus achieving a unified description of matter and the forces that govern its interactions. In particular, the graviton would have a corresponding gravitino.

As aesthetically pleasing as the theory was to many physicists, it failed to provide any experimental evidence in its defense. None of the super-symmetric partner particles has ever been discovered. Part of the difficulty is that although in theory the supersymmetric partners are the same mass as their normal companions, in our observable world, this symmetry is broken and the supersymmetric partners are hundreds or thousands of times heavier (and therefore require tremendous energy to discover in the lab). The hypothetical particle responsible for this broken symmetry is the Higgs particle, a spin 0 scalar particle that plays an important role in inflation theory.[36] Nevertheless, Hawking had stated in his Lucasian inaugural speech that he felt supergravity was the best candidate known for a complete unified theory. He had also been one of the hosts of a Nuffield Workshop on supergravity in the summer of 1980 and had coedited the conference proceedings.[37]

Former Cambridge student Nick Warner praised Hawking for his open-mindedness in these matters and the fact that it "was possible—indeed encouraged—for me as a student to move from my first work in relativity and Euclidean quantum gravity to the study of... supergravity."[38] Hawking's open-mindedness concerning his personal life was about to face a serious challenge.

NOTES

1. Jane Hawking, *Music to Move the Stars: A Life with Stephen Hawking* (London: Pan Books, 2000), p. 378.

2. Elizabeth Grice, "Dad's Important, But We Matter, Too," *The Telegraph*, April 13, 2004, http://www.telegraph.co.uk/arts/main.jhtml?xml=/arts/2004/04/13/bohawk13.xml (accessed May 25, 2004).

3. Richard Jerome, Vickie Bane, and Terry Smith, "Of a Mind to Marry," *People* (August 7, 1995): 46.

4. Emine Saner, "A Brief History of Mine," *The Scotsman*, April 20, 2004, http://lifestyle.scotsman.com/living/headlines_specific.cfm?articleid=8344 (accessed May 25, 2004).

5. Grice, "Dad's Important, But We Matter, Too."

6. Stephen Hawking, *Black Holes and Baby Universes and Other Essays* (New York: Bantam Books, 1993), p. 68.

7. Stephen Hawking, ed., *Stephen Hawking's A Brief History of Time: A Reader's Companion* (New York: Bantam Books, 1992), pp. 151–52.

8. Hawking, *Music to Move the Stars*, pp. 409–12.

9. H. A. Brück, "Introductory Remarks," in *Astrophysical Cosmology*, ed. H. A. Brück, G. V. Coyne, and M. S. Longair (Vatican City: Pontificia Academia Scientiarum, 1982), p. xxxiv.

10. S. W. Hawking, "The Boundary Conditions of the Universe," in *Astrophysical*

Cosmology, ed. H. A. Brück, G. V. Coyne, and M. S. Longair (Vatican City: Pontificia Academia Scientiarum, 1982), p. 571.

11. John Gribbin, *In Search of the Big Bang* (Toronto: Bantam Books, 1986), p. 388.

12. Hawking, ed., *Stephen Hawking's A Brief History of Time*, p. 120.

13. The problems with the big bang model are described in Appendix C.

14. The strong nuclear force "glues" quarks together to form protons and neutrons, while the weak nuclear force governs radioactive decay.

15. S. W. Hawking, "Perturbations of an Expanding Universe," *Astrophysical Journal* 145 (1966): 544.

16. S. W. Hawking, I. G. Moss, and J. M. Stewart, "Bubble Collisions in the Very Early Universe," *Physical Review D* 26 (1982): 2692.

17. American physicists Andreas Albrecht and Paul Steinhardt of the University of Pennsylvania developed their own "new inflation" independently of Linde and at approximately the same time. All three men are now generally given credit for the idea.

18. Andrei Linde, "Inflationary Theory Versus the Ekpyrotic/Cyclic Scenario," in *The Future of Theoretical Physics and Cosmology*, ed. G. W. Gibbons, E. P. S. Shellard, and S. J. Rankin (Cambridge: Cambridge University Press, 2003), p. 802.

19. Hawking, *Music to Move the Stars*, pp. 402–403.

20. A. D. Linde, "A New Inflationary Universe Scenario: A Possible Solution of the Horizon, Flatness, Homogeneity, Isotropy, and Primordial Monopole Problems," *Physics Letters* B108 (1982): 389–93.

21. Stephen Hawking, ed., *Stephen Hawking's A Brief History of Time* (New York: Bantam Books, 1988), p. 131.

22. S. W. Hawking and I. G. Moss, "Supercooled Phase Transitions in the Very Early Universe," *Physics Letters* B110 (1982): 35.

23. S. W. Hawking, "The Development of Irregularities in a Single Bubble Inflationary Universe," *Physics Letters* B115 (1982): 295–97.

24. Alan Guth, "Inflation and Cosmological Perturbations," in *The Future of Theoretical Physics and Cosmology*, ed. G. W. Gibbons, E. P. S. Shellard, and S. J. Rankin (Cambridge: Cambridge University Press, 2003), p. 725.

25. G. W. Gibbons, S. W. Hawking, and S. T. C. Siklos, ed., *The Very Early Universe* (Cambridge: Cambridge University Press, 1983).

26. Hawking, *Music to Move the Stars*, p. 382.

27. Hawking, *Black Holes and Baby Universes and Other Essays*, p. 94.

28. Ibid., p. 81.

29. Stephen Hawking, *The Universe in a Nutshell* (New York: Bantam Books, 2001), p. 85.

30. S. W. Hawking, "Quantum Cosmology," in *Three Hundred Years of Gravitation*, ed. S. W. Hawking and W. Israel (Cambridge: Cambridge University Press, 1987), p. 651.

31. J. B. Hartle and S. W. Hawking, "The Wave Function of the Universe," *Physical Review D* 28 (1983): 2975.

32. Hawking, *Black Holes and Baby Universes and Other Essays*, p. 81.

33. Stephen W. Hawking, *The Theory of Everything* (Beverly Hills: New Millennium Press, 2002), p. 118.

34. The SLAC SPIRES-HEP all-time most-cited list can be found at http://www.slac.stanford.edu/library/topcites/2003/eprints/gr-qc.topcites.all.shtml.

35. Strictly speaking, protons and neutrons are not fundamental particles but are instead composed of quarks. Quarks are fermions, so the statement that fermions make up matter still holds true.

36. Refer to Appendix C for a detailed discussion of symmetry breaking and the Higgs boson.

37. S. W. Hawking and M. Rocek, ed., *Superspace and Supergravity* (Cambridge: Cambridge University Press, 1981).

38. Nick Warner, "Gauged Supergravity and Holographic Field Theory," in *The Future of Theoretical Physics and Cosmology*, ed. G. W. Gibbons, E. P. S. Shellard, and S. J. Rankin (Cambridge: Cambridge University Press, 2003), p. 494.

CHAPTER 8

CHALLENGES AND CONTROVERSY

An Unexpected Silence and Time's Arrows

A BRUSH WITH DEATH

In the spring of 1982, Stephen gave three Morris Loeb Lectures at Harvard on various aspects of gravitational collapse. This experience led him to ponder the possibility of writing a popular-level book on his research. He was certain that "nearly everyone was interested in how the universe operates, but most people cannot follow mathematical equations—I don't care much for equations myself. This is partly because it is difficult for me to write them down."[1] There was a less altruistic, albeit more realistic, reason for writing such a book. Lucy was now eleven years old and completing grade school. Robert had benefited from private school, and the same was certain to be true for Lucy. However, in order to send Lucy to the Perse School without straining their savings, a new source of revenue had to be found. With these two motivations in mind, Hawking began working on the manuscript, completing the first draft in 1984.

Hawking's previous books had all been published by the prestigious Cambridge University Press. This was no academic tome, though, and Hawking wanted a publisher more accustomed with marketing popular works. He contacted Al Zuckerman, a literary agent, the brother-in-law of a colleague, who passed the first draft to several publishers. Hawking accepted an offer from

Bantam Books, a publisher with a solid foothold in the popular market, and awaited the customary give-and-take process certain to occur with his new editor. Secretarial duties had meanwhile transferred from Judy Fella, who left her position in order to accompany her husband on an extended trip to South Africa, to Laura Gentry.[2]

Hawking's fame continued to grow in 1985, now fueled by his first book-length biography, written by science journalist John Boslough.[3] He was also awarded the Gold Medal of the Royal Astronomical Society, its highest honor, for achievement in astronomy. Around this time, Hawking's personal life had also taken an important turn. Jane confided to him that her relationship with Jonathan had matured from friendship to romance. Nevertheless, she had no wish to disrupt their family. She later described how

> Stephen said he would not object so long as I continued to love him. I could not fail to love him when he showed such understanding.... At the time I felt very guilty but [Jonathan] was a godsend. We were rarely alone together and tried to maintain our code of conduct in front of Stephen and the children, suppressing displays of close affection.... Jonathan and I had struggled with our own consciences and had decided that the greater good—the survival of the family unit, Stephen's right to live at home within that family unit and the welfare of the children—outweighed the importance of our relationship.[4]

Indeed, all involved were so discreet that no one outside their circle of family and friends knew of this unconventional arrangement.

Summer found a grand adventure planned for the extended family. Stephen, Laura Gentry, and several students and nurses, were to spend a month in Geneva at the Conseil European pour la Recherché Nucleaire (CERN) particle accelerator facility, while the rest of the family enjoyed the sights of Lake Geneva. The only exception was Robert, who was traveling to Iceland for a Venture Scouts expedition. Stephen headed off first, leaving Jane, Jonathan, and the two younger children to camp their way across Belgium and Germany before meeting up with him and his entourage at Bayreuth for a production of Wagner's Ring Cycle. At a campground in Mannheim, Germany, Jane stopped at a public phone to check in with Stephen, and to her shock, Laura explained that Stephen was seriously ill and in the hospital on life support.

The family rushed to Stephen's side, and learned that the lingering cough he had picked up on a trip to China had quickly developed into pneumonia.

He was in a drug-induced coma and breathing through a respirator. Jane fell into "a black pit of anxiety, misery, and guilt. How could I have ever let Stephen go off alone with his entourage, deprived of my intimate knowledge of his condition, his needs, medicines, dislikes, allergies, and fears?"[5] The physician, not understanding Stephen's will to live despite his deteriorated physical condition, asked Jane if she wished for him to be disconnected from life support and allowed to die. Horrified, Jane insisted that no such thing would be done. The physician then explained the future that Stephen now faced. He would have to undergo a tracheotomy, where a permanent hole would be cut in his trachea below his vocal cords. On the positive side, his coughing fits would cease, as the highly sensitive area of his throat would be bypassed. On the other hand, the operation would permanently rob him of what little garbled speech he had left, and require constant monitoring for the rest of his life. In other words, Stephen would have to move from nursing care several hours a day to round-the-clock care.

When Stephen recovered some of his strength, the university paid for an air ambulance to fly him back to Cambridge, where he was admitted to Addenbrooke's Hospital. Judy Fella visited Stephen in the intensive care unit, and she immediately began helping Jane make arrangements for Stephen's eventual return home. Although valiant attempts were made to try and wean him off the respirator naturally, the choking fits returned, and performing the tracheotomy became unavoidable. It was during these desperate times that Hawking had vivid dreams of flying in a hot-air balloon, which to him was a symbol of hope.[6] The Hawking family received a more tangible reason for hope, as Robert had received good grades on his A-level exams and would be accepted into Cambridge for the following fall to major in science.

The operation was a success, and Stephen began to regain more of his strength. He was now in imminent danger of being locked within his own closed universe; unable to speak or write, how was he to communicate with the outside world? Initially he was only able to painstakingly spell out words one letter at a time by raising his eyebrows when someone pointed to the correct letter on an alphabet card. As he later described, it is "pretty difficult to carry on a conversation like that, let alone write a scientific paper."[7]

IN A DIFFERENT VOICE: COMPUTERS TO THE RESCUE

Walt Woltosz, a computer expert from California, sent Stephen a program he had developed in order to help his disabled mother-in-law communicate. It not only selected words from a dictionary menu, but it also had a built-in speech synthesizer. Originally it used electrodes attached to the head to select words displayed on a computer screen, but one of Hawking's students adapted the device to a handheld controller, similar to a computer mouse, allowing him to click on words and phrases to compose sentences. For the first time in many years, Hawking was now able to write, and speak in a clear, albeit emotionless, voice.

With the communication problem solved, there remained one more, equally large, challenge. The tracheostomy—the actual permanent hole in his windpipe—would require constant attention by trained medical professionals. The tube that was now inserted in his throat to help him breathe had to be "cleaned regularly by a sort of mini-vacuum cleaner to bring up the secretions which perpetually accumulated in his lungs, and the device itself was potentially a source of damage and dangerous infection."[8] He would have to have nursing care around the clock. Where was the money going to come from? Martin Rees' generous donors had covered part-time care, but funding this new level of treatment would tap someone's very deep pockets.

Judy Fella had alerted Stephen's colleagues to the dire situation, and one of them immediately came to the rescue. Kip Thorne contacted Jane, suggesting that she seek funding from the famous John D. and Catherine T. MacArthur Foundation, one of the largest philanthropic organizations in the United States. Among their board members was Murray Gell-Mann, a famous particle physicist. The foundation granted Jane's request on a trial basis, providing just enough money to cover constant nursing care. She and Laura immediately set to work finding nurses. Stephen began to come home on Sunday afternoons, and his release was finally set for November 4, more than three months after his initial hospitalization.

With a great deal of hard work, Laura Gentry and Jane managed to put together a host of suitable nurses to cover three shifts per day, some from agencies and others privately hired. Many decided the job was not to their liking, while others genuinely tried to adapt to the demanding position only to reluctantly quit later after simply burning out.[9] Eventually a core of dedicated, long-term nurses emerged, among them Elaine Mason. The freckled redhead had two sons of her own, one the same age as Tim Hawking and the other two years younger. Her husband, David, was a computer engineer, and he ingeniously found a way to mount Stephen's computer, screen, and speech synthesizer to the electric wheelchair, allowing him to communicate without being chained to a desk.

While having the nurses allowed Stephen to live at home once more, the situation brought a new source of tension to the family. Privacy was nonexistent, not only between Jane and Stephen, but for the children as well. This was especially hard on Robert and Lucy, both of whom were now teenagers. It was at this time that Laura Gentry, who had been a great source of support and aid to the Hawkings, left her position, but fortunately Judy Fella agreed to resume her duties.

Stephen was able to return to work before Christmas, initially on a part-time basis, and always accompanied by a nurse. The new year found him in good spirits with an optimistic outlook. He was able to resume his research in physics and collaborate with students and colleagues much more easily than before. Although he had lost his natural voice, it had already become so slurred and indistinct that only a handful of people could understand it. Public appearances had become public readings, where one of his students would read his prepared talk, while he sat there as little more than set decoration. He found that with his speech synthesizer he could be "a successful public speaker, addressing large audiences. I enjoy explaining science and answering questions."[10] The process of composing sentences one selected word at a time was painstakingly slow, but he didn't fall into the habit of leaving out small words such as "the." He did learn to become quite succinct in his statements, saying as much as possible in relatively few words. This ability to transmit information as punch lines certainly fit his sharp sense of humor. In fact his only (often repeated) joking complaint about his communication system was that it gave him an American accent. His daughter Lucy later described him as "incredibly tenacious; a very stubborn person. Once he sets his mind on something, he's like an ocean liner, he doesn't change course. He's defied the disease. He's defied perceptions of a disabled man. I think what he's done is amazing."[11]

Sadly, the positive atmosphere was tainted with tragedy in March 1986. Stephen's father, Frank, had been ill for some months and finally passed away. Isobel described Stephen as "very upset by his father's death—it was rather a dreadful thing.... He was very fond of his father, but they had grown apart rather and hadn't seen a great deal of each other in the late years."[12]

Stephen slowly resumed his travel schedule, first attending a conference in Sweden. There, Murray Gell-Mann was able to see firsthand how the MacArthur money was being used to help Hawking continue not only his home life but also his valuable scientific work. Jane referred to this meeting at her September application to the foundation, who agreed to extend their funding of Hawking's medical expenses on an ongoing basis.[13] In October the Hawkings traveled to the Vatican once again, as Stephen was appointed to the Pontifical Academy of Sciences. The family was granted an audience with Pope John Paul II, who extended his blessings.

THE DIRECTION OF TIME

Now that Stephen was back at work, he could resume some long-interrupted research projects. Chief among them was controversial work that had made a stir when the original paper was published the previous November—namely the problem of the direction or arrow of time in cosmology. Hawking had been introduced to the topic in graduate school by his advisor, Dennis Sciama. He did review the literature on the topic at the time, but decided he "needed something more definite and less airy-fairy" for his thesis topic, so he changed directions and began to work on singularity theorems, which he called "a lot easier."[14] His interest in the topic was reawakened during his work with Jim Hartle on the no boundary proposal.

The main question was as follows: why do the three "arrows of time"—psychological, thermodynamic, and cosmological—all seem to point in the same direction, and would this necessarily always be true throughout the evolution of the universe? The psychological arrow of time is the way time appears to flow to the human mind—the way we experience the unceasing aging of our bodies, remember the past, and have no direct knowledge of the future. The thermodynamic arrow of time refers to the fact that entropy increases with time (from the second law of thermodynamics). The cosmological arrow speaks to the evolution of the universe, currently expanding from the big bang. The connection between the psychological and thermodynamic arrows can be argued from an analogy with computers:

> When a computer records something in memory, the total entropy increases. Thus computers remember things in the direction of time in which entropy increases?... It seems reasonable to assume that we remember in the same

> direction of time that computers do.... This means that the psychological arrow of time, our subjective sense of time, is the same as the thermodynamic arrow of time, the direction in which entropy increases.... So the second law of thermodynamics is really a tautology. Entropy increases with time because we define the direction of time to be that in which entropy increases.[15]

What was the connection between the cosmological arrow of time and the others? In his 1985 paper, Hawking put forth the controversial conjecture that although the three arrows all point in the same direction during the current, expanding phase of the universe, he believed that the no-boundary proposal (which utilized closed universes doomed to eventually collapse in on themselves) would predict that when the universe began to contract, the thermodynamic arrow of time (and so by the previous argument, the psychological arrow of time as well) would reverse. He proposed that entropy would decrease rather than increase, meaning that things would become more ordered rather than more random. Humpty Dumpty would finally reassemble, and people would grow younger rather than older. We would remember the future and have no knowledge of the past.

Hawking thought this reversal of the arrows of time in the "contracting phase had a satisfactory ring to it."[16] However, others disagreed with him. Long-time colleague Don Page and student Raymond LaFlamme, using slightly different methods from each other, got results that openly conflicted with their mentor's. After about a month of discussions and shared calculations, they managed to convince Stephen that they were right. Hawking's paper had already been submitted and was being finalized for publication, and since it was only one (albeit important) aspect of the paper that was in question, the work was published as written, with an additional note tacked onto the end:

> Since this paper was submitted for publication a paper by Don Page has appeared.... In it he questions my conclusion that the thermodynamic arrow of time would reverse in a contracting phase of the universe or in a black hole.... I think that Page may well be right in his suggestion.[17]

The paper to which Hawking referred was published in the same issue as his, in the pages directly following his paper.[18]

Hawking had meant to explore the implications of LaFlamme and Page's calculations during his trip to CERN, but his illness and subsequent tracheotomy had put that all on hold. Now it was time to complete his reanalysis of the problem and announce his findings to the physics community. In December 1986 he attended the Texas Symposium of Relativistic Astrophysics (so-named because the conference had originated in Texas), held in Chicago. The local media treated him "as if he were visiting royalty or a rock star,"[19] hanging on each electronically spoken word. While in Chicago he gave a widely attended talk in which he formally announced that he had been wrong about the reversal of the psychological and thermodynamic arrows of time in a contracting universe. He later called the incident his "greatest mistake, or at least my greatest mistake in science. I once thought there ought to be a journal of recantations, in which scientists could admit their mistakes. But it might not have many contributors."[20]

WHY IS THE UNIVERSE THE WAY IT IS?

What was the final outcome of the controversy? Hawking's eventual argument made use of what is called the anthropic principle. He commonly defined it as the principle that "the universe has to be more or less as we see it, because if it were different, there wouldn't be anyone here to observe it."[21] Robert Dicke of Princeton suggested in 1961[22] that calculations of the age of the universe are limited by the fact that the universe must be old enough for observers (namely humanity) to exist to make the measurements in the first place. This is based on the fact that the first generation of stars created in the early universe was composed of nearly pure hydrogen and helium. Elements necessary for life as we know it—carbon, oxygen, and nitrogen—were created for the first time in the death throes of those earliest stars. As succeeding generations of stars were born and died, the chemistry of the universe continued to be enriched, leading to the abundances of the elements we currently observe. According to this scenario, neither carbon-based observers such as human beings, nor the earth itself, could exist until billions of years after the big bang. Other "fine-tunings" of the universe appear as well. If the strengths of the fundamental forces of nature were slightly different, there might be disastrous consequences. Atoms might be unstable, or stars might not live long enough for complex life forms to evolve on their planets. Hawking noted that "a universe like ours with galaxies and stars is actually quite unlikely. If one considers the possible constants and laws that could have emerged, the odds against a universe that has produced life like ours are immense."[23]

Hawking's colleague Brandon Carter had pondered those unlikely coincidences for several years and had come up with what he had named the anthropic principle; as he stated it, "what we can expect to observe must be restricted by the conditions necessary for our presence as observers."[24] Although Carter unveiled his idea in widely noted form at the Copernicus conference he and Stephen had attended in Poland in 1973, Hawking had already been well aware of his friend's inklings for several years. In fact, he first used and referenced Carter's theory in his own research in a paper published shortly before the Copernicus conference. Hawking and C. B. Collins connected the fact that the overall distribution of matter in the universe is remarkably similar in all directions we observe (i.e., it is isotropic) to the density perturbations in the early universe (which led to the creation of the galaxies), and found the balance to be highly unlikely. They came to the conclusion that "the fact that we have observed the universe to be isotropic is therefore only a consequence of our own existence"[25]—if it were otherwise, galaxies would not exist, and neither would theoretical physicists posing such questions.

With these concepts in mind, Hawking pondered the problem of the three arrows and came to the conclusion that although the psychological and thermodynamic arrows did not reverse during the contracting phase of the universe, it was only when the three arrows pointed in the same direction that intelligent life could exist who could contemplate the problem. Given observations of the rate of expansion of the universe, the universe could not begin to collapse for a very long time (if ever). This would be so far into the future that all the stars would be long dead, and therefore life would be impossible in the collapsing phase. Hawking stated, "the fact that we are around to observe the universe, means that we must be in the expanding, rather than the contracting phase."[26]

Although Stephen apparently had no problem invoking this principle to explain the enigmatic "fine-tunings" of the universe, many physicists considered the principle the equivalent of scientific voodoo. David Gross, a colleague of Jim Hartle's at the University of California, Santa Barbara, called it "dangerous" because it has been used by some to bolster the notion that science can provide evidence for the existence of a God who purposefully set up the universe in such a way as to allow for the eventual evolution of human beings. "It smells of religion... and like religion, it can't be disproved."[27] However, others have pointed out that there need not be anything religious about the idea, as it is just another example of nature rolling the dice. Gordon Kane of the University of Michigan compared it to a lottery, in that if you "win the lottery, you may feel very grateful, but someone had to win, and no one selected who that was, except randomly. Just because the universe has a unique set of laws and parameters should not lead one to wonder whether that set was designed."[28] This would not be the last time Hawking would invoke the anthropic principle, and certainly not his final brush with controversy. Previously, discussion and debate of these issues had been largely limited to the scientific arena, but Hawking was soon to introduce these mind-bending concepts to a new, and far-wider, audience.

NOTES

1. Stephen Hawking, *Black Holes and Baby Universes and Other Essays* (New York: Bantam Books, 1993), p. 35.

2. Jane Hawking, *Music to Move the Stars: A Life with Stephen Hawking* (London: Pan Books, 2000), p. 429.

3. John Boslough, *Stephen Hawking's Universe* (New York: Quill/William Morrow, 1985).

4. Lisa Sewards, "A Brief History of Our Life Together," *Daily Telegraph*, April 27, 2002, http://www.lexisnexis.com (accessed August 19, 2004).

5. Ibid.

6. Nigel Farndale, "A Brief History of the Future," *Sydney Morning Herald*, January 7, 2000, http://www.smh.au/news/0001/07/features/features1.html (accessed March 27, 2003).

7. Hawking, *Black Holes and Baby Universes and Other Essays*, p. 25.

8. Hawking, *Music to Move the Stars*, p. 443.

9. Ibid., pp. 452–57.

10. Stephen Hawking, *Black Holes and Baby Universes and Other Essays*, p. viii.

11. Elizabeth Grice, "Dad's Important, But We Matter, Too." *The Telegraph*, April 13, 2004, http://www.telegraph.co.uk/arts/main.jhtml?xml=/arts/2004/04/13/bohawk13.xml (accessed May 25, 2004).

12. Stephen Hawking, ed., *Stephen Hawking's A Brief History of Time: A Reader's Companion* (New York: Bantam Books, 1992), p. 161.

13. Hawking, *Music to Move the Stars*, p. 470.

14. S. W. Hawking, "The No-Boundary Proposal and the Arrow of Time," in *Physical Origins of Time Asymmetry*, ed. J.J. Halliwell, J. Perez-Mercader, and W. H. Zurek (Cambridge: Cambridge University Press, 1992), p. 268.

15. Ibid.

16. Ibid., p. 271.

17. S. W. Hawking, "The Arrow of Time in Cosmology," *Physical Review D* 32 (1985): 2495.

18. Don N. Page, "Will Entropy Decrease if the Universe Recollapses?" *Physical Review D* 32 (1985): 2496–99.

19. John Boslough, *Masters of Time* (Reading, MA: Addison-Wesley, 1992), p. 4.

20. Hawking, "The No-Boundary Proposal," p. 275.

21. Stephen W. Hawking, *The Universe in a Nutshell* (New York: Bantam Books, 2001), pp. 85–87.

22. R. H. Dicke, "Dirac's Cosmology and Mach's Principle," *Nature* 192 (1961): 440–41.

23. Boslough, *Stephen Hawking's Universe*, p. 123.

24. Brandon Carter, "Large Number Coincidences and the Anthropic Principle in Cosmology," in *Modern Cosmology and Philosophy*, ed. John Leslie (Amherst, NY: Prometheus Books, 1998), p. 132.

25. C. B. Collins and S. W. Hawking, "Why is the Universe Isotropic?" *Astrophysical Journal* 180 (1973): 319.

26. Hawking, "The No-Boundary Proposal," p. 274.

27. Dan Falk, "The Anthropic Principle's Surprising Resurgence," *Sky and Telescope* (March 2004): 46.

28. Gordon Kane, "Anthropic Questions," *Phi Kappa Phi Forum* 82, Fall (2002): 24.

CHAPTER 9

THE BEST SELLING BOOK THAT "NO ONE READ"

A Brief History of Time

PRIZES AND PUBLISHERS

After his triumphant return to the world of physics, Hawking was ready to resume working toward his goal of writing a popular-level work on cosmology. The first draft of his manuscript had been accepted by Bantam Books in the summer of 1985, and his new editor, Peter Guzzardi, had prepared pages of critiques and suggestions. However, Stephen's illness had left him in no position to work on the draft, and it languished in limbo for many months. With his new communication program, and the considerable help of student Brian Whitt, he was slowly able to make the major required revisions. Although he admitted being rather irritated at Guzzardi's lengthy list of necessary changes, he also came to realize that his editor was right. Hawking commented that "it is a better book as a result of his keeping my nose to the grindstone."[1] The second draft was completed in the spring of 1987, and despite Guzzardi's warning that every equation would cut the sales in half, Hawking could not resist including perhaps the most famous scientific formula of all time—Einstein's $E = mc^2$.

With his first popular work now well on its way to publication, Hawking returned to more scholarly works. Cambridge held an international conference in late June–early July 1987 to celebrate the tercentenary of the publication of

Sir Isaac Newton's most famous work, *Principia Mathematica.* As Lucasian Professor, Hawking was considered in some ways a successor of Newton, and he was instrumental in the planning of the event. He and colleague Werner Israel solicited articles on various aspects of gravitational theory, which they edited into a volume in part to "supplement and update the articles in a book... edited previously to mark the centenary of Albert Einstein's birth."[2] The same year, Hawking received the first Paul Dirac Medal from the Institute of Physics, for his outstanding contributions in theoretical physics.

The following spring Jane and Stephen flew to Jerusalem where he and old friend Roger Penrose were jointly awarded the Wolf Prize in Physics. Conferred by the Wolf Foundation of Israel, it is recognized as second only to the Nobel Prize in Physics in its prestige. President Chaim Herzog presented the award to Hawking and Penrose, citing them for their "brilliant development of the theory of general relativity, in which they have shown the necessity for cosmological singularities and have elucidated the physics of black holes."[3] In his acceptance speech, Hawking took the opportunity to speak for peace:

> The progress of science has shown us that we are a very small part of the vast universe, which is governed by rational laws. It is to be hoped that we can also govern our affairs by rational laws, but the same scientific progress threatens to destroy us as all?... Let us do all we can to promote peace and so insure that we will survive till the next century and beyond.[4]

While in Israel, Stephen was asked by a reporter about his religious beliefs. Stephen's answer was consistent with his previous comments on the issue, and he said that he "did not believe in God and there was no room for God in his universe." Jane felt bitter at Stephen's open denial of everything she believed in, especially as spoken in one of the holiest cities in the world. It had been her faith that had sustained her in her marriage—faith in Stephen's courage and genius, faith in their joint efforts, and religious faith. She found Stephen's ill-defined atheism yet another wedge, which threatened to drive them farther apart.[5]

Advanced copies of Stephen's book, *A Brief History of Time*, were now circulating. An eagle-eyed reviewer for the journal *Nature* found copious, embarrassing errors, such as mislabeled photographs and figures. In response, Bantam immediately recalled the entire printing and hurriedly made the corrections in order for the book to meet its April release date.[6] The book was launched at a private luncheon at the Royal Society of London on June 16, a week after Jane was stricken with a painful case of shingles. Nevertheless she was proud of her husband and his achievement, saying that it "involved us both in a way that was reminiscent of those passionate struggles and heady victories in the early years of our marriage."[7]

The book soon became an international sensation, beyond Stephen's wildest dreams. It remained on the *New York Times* bestseller list for fifty-three weeks, and the London *Sunday Times* bestseller list for over four years. This last feat landed the book in the *Guinness Book of Records*. It was eventually translated into over sixty languages. The only problem was that it had the reputation of being nearly unreadable. It was rumored that people bought it to keep on their coffee tables as a conversation piece. While this is certainly an exaggeration, Hawking himself later admitted that the book was not easygoing. He reflected that he did not explain the sum over histories method and imaginary time as fully as he should have. He especially noted that imaginary time was "the thing in the book with which people have the most trouble. However, it is not really necessary to understand exactly what imaginary time is—just that it is different from what we call real time."[8]

Why was the book such a phenomenal success? Did people buy it because of positive reviews or its status on the bestseller list? Hawking suggested that some of the sales might have been due to people's interest in him as a disabled person. Certainly the cover photograph of Hawking in his electric wheelchair juxtaposed against a starry background was eye-catching. Some accused Bantam of exploiting Hawking's disability by using the picture. Readers who bought the book thinking it was a biography would have been immediately disappointed, as Stephen said he had written it "as a history of the universe, not of me."[9] The public's fascination with him had already been primed by an article on black holes in *Time* that February, which included a special sidebar feature on Hawking. In it, Rocky Kolb of Fermilab compared Hawking to basketball star Michael Jordan, saying that "No one can tell Jordan what moves to make. It's intuition. It's feeling. Hawking has a remarkable amount of intuition."[10] For his part, Hawking acknowledged that his readers might not have understood the entire book, but said that "they have got a feeling of being in touch with the big questions: Where did we come from, and how did it all begin?"[11]

Despite its reputation as a formidable text, readers were immediately drawn in by the introduction, written by famed American popularizer of science, Carl Sagan. He described the work as "a book about God... or perhaps about the absence of God," drawing attention to Hawking's scientific goal of understanding the mind of God. Sagan also wrote that this understanding "makes all the more unexpected the conclusion of the effort, at least so far: a universe with no edge in space, no beginning or end in time, and nothing for a Creator to do."[12] Those who read the book to the end (or cheated, and snuck a peek at the last page) found a controversial passage destined to become an often repeated and cited quotation, that claimed that when and if the true unified theory in physics is discovered,

> we shall all, philosophers, scientists, and just ordinary people, be able to take part in the discussion of the question of why it is that we and the universe exist. If we find the answer to that, it would be the ultimate triumph of human reason—for then we would know the mind of God.[13]

WORMHOLES AND BABY UNIVERSES

While *A Brief History of Time* was taking on a life of its own, the Hawking family was going forward with their lives. Lucy had a part in the Cambridge Youth Theatre's award-winning performance of *The Heart of a Dog*, a 1920s Soviet political satire. Unfortunately, their London run conflicted with the entrance exam for Oxford, and Lucy had to change her university application to take into account that she would miss the entrance exam and instead would be judged solely on an interview and her A-level exams.[14] Stephen attended his daughter's opening night and then flew with his nurses to California for a month. He had been awarded the Hitchcock Lectureship by the University of California, Berkeley, and while there he gave three public lectures, "The Origin of the Universe," "The Direction of Time," and "Black Holes, White Holes, and Worm Holes." The third of these was connected with Hawking's latest provocative research topic.

What happens to particles that fall into a black hole? Hawking considered what his no-boundary proposal predicted in this situation. He realized that since the class of universe models he was studying had no edges or singularities, there had to be somewhere for the particles that fell into the hole to go. He reasoned that a "wormhole leading off to another region of space-time, would seem to be the most reasonable possibility."[15] Wormholes can be thought of as tunnels that represent shortcuts in space and time. They are one of several convenient devices in science fiction (such as in *Babylon 5* and *Contact*) when a writer doesn't want to deal with the limitations of the speed of light. It is more dramatic to have the crew travel faster than light (as in *Star Trek*, with their so-called warp speed) or travel through hyperspace (as in *Star Wars*) or through a wormhole.

Hawking had a slightly different version of his theory as well. Instead of the wormhole leading to another arbitrary region of space-time, it could also lead to an independent, closed universe. This would be a baby universe. How many baby universes could there be? Hawking noted that they "exist in a realm of their own. It is a bit like asking how many angels can dance on the head of a pin."[16] These baby universes became central to his controversial idea concerning information loss in black holes. The particles that went down the drain into the baby universes would be information lost to our universe.

If Hawking's wormholes and baby universes existed, could they be used as an intergalactic or interuniversal transportation system, as in science fiction? Sadly, the answer was a resounding no. The problem was that because Hawking was using his no-boundary technique, he was doing calculations in imaginary time. In real time, the singularity at the center of the black hole would still exist, and anything entering the black hole would suffer the unfortunate fate of being torn to shreds by extreme gravitational effects. Hawking could only offer the rather unsatisfying consolation prize of having the history of the black hole's meal continue in imaginary time. He said that "it would pass into the baby universe and would reemerge as the particles emitted by another black hole" in that baby universe. His tongue-in-cheek advice for anyone falling into a black hole was to "think imaginary."[17]

While Stephen was pondering imaginary wormholes, Jane was considering more pragmatic issues. With the money from the Wolf Prize and the royalties from Stephen's book, they could consider buying a second home. Stephen wanted to buy a property in Cambridge as an investment, but Jane had dreams of buying an old farmhouse in France and renovating it. While Stephen was in Berkeley, Jane and Timmy traveled to France and fell in love with an old millhouse that Jane named "The Moulin." The property was purchased in March 1989, and Jane quickly set to work on renovations. It would become her sanctuary in the personal storm that was about to break around them.[18]

PHYSICIST AS CELEBRITY

With the worldwide success of his book, Hawking became not only a household name but also a bona fide celebrity. In October 1988, Jane and Timmy joined Stephen in Barcelona, Spain, for the launch of the Spanish edition of the work. Jane said that he was recognized everywhere, "attracting crowds who stopped to applaud him in the street. So much sudden attention was gratifying and disturbing at one and the same time."[19] This instant notoriety intruded upon the Hawkings' home as well. Would-be physicists called at all hours of the day and night, expecting to be able to talk directly to Hawking. One caller went so far as to propose marriage to Lucy on the condition that her father read his thesis.[20] Hawking was named one of the "Twenty-Five Most Intriguing People" of 1988 by *People* magazine. In what was supposed to be a tribute, the editors pointed out that "what's left of Stephen Hawking, the physical man, is a big head ripped by a drooling grin and a body collapsed into a pile of wasted limbs."[21] This unusually crass description articulated how some people actually saw Hawking—as something of a freak show, rather than a talented human being who happened to be challenged by physical disability.

Fame also had its amusing and inspiring aspects. "Bloom County" comic strip author, Berke Breathed, created an imaginary off-screen rivalry to develop the theory of everything between Hawking and his child-genius character, Oliver Wendell Jones. Susan Anderson and Bill Allen, owners of the Gold Star Sardine Bar in Chicago, printed and distributed free Stephen Hawking fan club T-shirts. Faced with overwhelming demand, they gave away eight thousand shirts in two months, including several to Hawking at the professor's amused request. Their motivation was to honor what they called a "real hero."[22] Hawking was even interviewed by *Playboy* magazine in the late summer of 1989.

In addition to the anointing that Hawking had received in the popular press, accolades and awards came from official avenues. He was given an honorary doctorate in Science from the University of Cambridge (conferred by the Duke of Edinburgh himself) in the summer of 1989. Soon thereafter, Queen Elizabeth presented Stephen with a medal for Companion of Honour, a title that ranked above that of knighthood. He was also a recipient of a Britannica Award for being a "supreme communicator to lay readers of extremely difficult physical concepts."[23]

Hawking appeared to shrug off the constant comparisons in the press between himself and Albert Einstein, and said that such people "don't understand either Einstein's work or mine."[24] Jane, however, was admittedly affected by the media onslaught, and said that she believed that in their eyes she was

> an appendage, a peep-show, relevant to Stephen's survival and his success only insomuch as in the distant past I had married him, made a home for him and produced his three children. Nowadays I was there to appease the media's desire for comforting personal detail by performing like a well-behaved circus animal.[25]

The constant media attention was a serious distraction to the older children, as Lucy studied for her A-level exams (upon which her entrance to Oxford hinged) and Robert was busy with final exams. Lucy later commented that although they weren't a normal family, they had achieved a "precarious balance. But then a combination of factors—the nurses, sudden fame, traveling a lot more—changed the dynamic in the household. Fame would change anybody and it did change him."[26] The stress extended to Hawking's department as well, leading to the resignation of Judy Fella and his graduate assistant, Nick Phillips.

Jane actively fought back against the unrealistic spin the press was attempting to put on their lives, and began to be (sometimes brutally) honest about their struggles in interviews. She even commented to reporters that her role in Stephen's life had become "telling him that he was not God."[27] She said that perpetuating the false image of "cheerful self-sufficiency... would be cheating the many disabled people and their families who were probably suffering all the heartache, anxieties, privations, stresses, and strains that we ourselves had undergone in earlier years." She worried that others would point accusingly at the disabled and their families and wonder why they couldn't live up to the heroism of the Hawkings.[28] Indeed, biographer Kitty Ferguson warned, "few people will find a Jane Hawking. Few have Hawking's powers of concentration and self-control. Few have his genius."[29]

A SHOCKING SEPARATION

The public's false picture of the Hawkings as the perfect family was about to be forever shattered. Stephen had become romantically involved with Elaine

Mason, still one of his nurses. Jane was prepared to accept their relationship the same way Stephen had earlier accepted hers and Jonathan's, with the provision that it was discreet, posed no threat to the family, the children, or their home and that it did not negate her relationship with Stephen."[30] In October 1989, Lucy left for Oxford to study French and Russian; Robert left for postgraduate studies in Glasgow; and Stephen presented Jane with a letter of intent to leave her and move in with Elaine.[31] The actual separation waited until the following February, several months shy of their 25th wedding anniversary.

Due to the remarkable discretion of all concerned, the press was unaware of the breakdown of the Hawkings' marriage until the summer. The predictable feeding frenzy occurred, with Jane later describing the reporters "clustering round the gate, like a pack of baying hounds.... We were being hunted."[32] Physicists around the world were saddened and puzzled at the turn of events, as many of them had bought into the idyllic facade that had been presented for years. Rumors of infidelity widely circulated, as did the more palatable excuse that their split had been over religious differences. The problem was exactly that which Jane had dreaded—people had come to have an unrealistic picture of their family. Why else would it come as such a shock? Divorce and infidelity touch many marriages, even those as apparently long-lived as theirs. Studies have shown that divorce rates among the disabled are even higher than among the general population.[33] But the Hawkings had become icons, and were assumed to be above the failures of mere mortals.

Yet in truth, the Hawkings' marriage had been far from a failure. It had survived more than twenty years, longer than anyone had predicted, had produced three talented children, and had sustained Stephen through the struggles in his early career and with his increasing physical challenges. Hawking called himself lucky in most aspects of his life. He acknowledged in the introduction to *A Brief History of Time* that the "help and support I received from my wife, Jane, and my children, Robert, Lucy, and Timmy, have made it possible for me to lead a fairly normal life and to have a successful career."[34]

The breakup of Hawking's marriage also had no impact on his iconic status and his effectiveness as an advocate for the rights of the disabled. The University of Bristol set up a special dormitory for handicapped students aptly named "Hawking House," and a hands-on science museum was named after him in San Salvador. The honor was bestowed in recognition of his "fortitude in dealing with adversity."[35] Hawking publicly stated that it is "very important that disabled children should be helped to blend in with others of the same age. It determines their self-image.... How can one feel a member of the human race, if one is set apart from an early age. It is a form of apartheid."[36] In lecture after lecture, Hawking flatly stated that his greatest achievement was being alive and that he was happier now than when his condition first appeared.[37]

While closing some doors to his past, there was a small unfinished detail that he felt it was time to settle. His bet with Kip Thorne concerning whether or not Cygnus X-1 was a black hole had remained open for over 15 years. Kip's wife and parents had been aghast at the stakes (namely Kip's subscription to *Penthouse*), but odds that he would actually receive the magazine seemed low, as new evidence was trickling in concerning Cygnus X-1. In June 1990 while Kip was away in Moscow, Stephen and his entourage broke into Kip's office at Caltech, "found the framed bet, and wrote a concessionary note on it with validation by Stephen's thumbprint."[38] One bet may have been settled, but several more faced Hawking in the near future.

NOTES

1. Stephen W. Hawking, *A Brief History of Time* (Toronto: Bantam Books, 1988), p. vii.
2. S. W. Hawking and W. Israel, ed., *300 Years of Gravitation* (Cambridge: Cambridge University Press, 1987), p. xi.
3. Matthew Siegel, "Wolf Foundation Honors Hawking and Penrose for Work on Relativity," *Physics Today* (January 1989): 97.
4. Lawrence Zalcman, "Mathematicians Sweep 1988 Wolf Prizes," *The Mathematical Intelligencer* 11, no. 2 (1989): 46.
5. Jane Hawking, *Music to Move the Stars: A Life with Stephen Hawking* (London: Pan Books, 2000), pp. 494–95.
6. Stephen Hawking, *Black Holes and Baby Universes and Other Essays* (New York: Bantam Books, 1993), p. 36.
7. Hawking, *A Brief History of Time*, p. 499.
8. Hawking, *Black Holes and Baby Universes and Other Essays*, p. 36.
9. Ibid., p. 38.
10. Leon Jaraoff, "Roaming the Cosmos," *Time* (February 8, 1988): 60.
11. Stephen Hawking, interview by Larry King, *Larry King Weekend*, Cable News Network, December 25, 1999.
12. Carl Sagan, "Introduction," in Hawking, *A Brief History of Time*, p. x.

13. Hawking, *A Brief History of Time*, p. 175.

14. Hawking, *Music to Move the Stars*, pp. 503–504.

15. S. W. Hawking, "Do Wormholes Fix the Constants of Nature?" *Nuclear Physics* B335 (1990): 286.

16. Hawking, *Black Holes and Baby Universes and Other Essays*, p. 124.

17. Ibid., p. 122.

18. Jane Hawking, *Music to Move the Stars*, pp. 509–14.

19. Ibid., p. 504.

20. Ibid.

21. *People*, "Stephen Hawking" (December 26, 1988): 99.

22. *People*, "Suiting Science to a T(Shirt), Two Chicago Bar Owners Set Up a Stephen Hawking Fan Club" (September 11, 1989): 111.

23. Arthur Fisher, "Introduction," in "A Brief History of A Brief History," Stephen Hawking, *Popular Science* (August 1989): 70.

24. Stephen Hawking, "Questions and Answers," *Professor Stephen Hawking's Web Pages*, http://www.hawking.org.uk/about/qa.html (accessed September 14, 2004).

25. Hawking, *Music to Move the Stars*, pp. 525–26.

26. Emine Saner, "A Brief History of Mine," *The Scotsman*, April 20, 2004, http://lifestyle.scotsman.com/living/headlines_specific.cfm?articleid=8344 (accessed May 25, 2004).

27. Tim Adams, "Brief History of a First Wife," *The Observer*, April 4, 2004, http://observer.guardian.co.uk/review/story/0,,1185067.html (accessed August 27, 2004).

28. Hawking, *Music to Move the Stars*, p. 536.

29. Kitty Ferguson, *Stephen Hawking: Quest for a Theory of Everything* (New York: Franklin Watts, 1991), p. 138.

30. Hawking, *Music to Move the Stars*, p. 555.

31. Ibid., pp. 560–62.

32. Ibid., p. 574.

33. Peter T. Kilborn, "Disabled Spouses Are Increasingly Forced to Go It Alone," *New York Times* (May 31, 1999): sec. A.

34. Hawking, *A Brief History of Time*, p. vi.

35. Oliver Komar and Linda Buechner, "The Stephen Hawking Science Museum in San Salvador," *Journal of College Science Teaching* 30, no. 2 (2000): 144–45.

36. Bob Sipchen, "Simply Human: Wheelchair-bound Physicist Stephen Hawking Resists Efforts to Deify His Life or His Disabilities," *Los Angeles Times* (June 6, 1990): sec. E.

37. Arthur Fisher, "Master of the Universe," *Popular Science* (November 1999): 87.

38. Kip S. Thorne, *Black Holes and Time Warps* (New York: W. W. Norton and Co., 1994), p. 315.

CHAPTER 10

TO BOLDLY GO

Time Travel and Television

MAKING THE UNIVERSE SAFE FOR HISTORIANS

It was a rainy evening, March 5, 1991, shortly before 11 PM, and Hawking was returning to his apartment for the evening. He crossed Grange Road, with his attending nurse dutifully following behind. A car was approaching, but he estimated that he had plenty of time to make it safely to the other side. He was wrong. The speeding taxi struck him from behind, throwing him into the road. His wheelchair was critically damaged, as was his computer system. He suffered a broken upper arm and cuts on his head serious enough to require stitches. Despite the injuries to his body and his damaged wheelchair, he was out of the hospital and back to work two days later.[1] The event was widely reported in the press, his brush with death only adding to his legend.

Hawking's reputation as a researcher of controversial theories continued to grow as well. While writing his novel *Contact*, Carl Sagan had sent the manuscript to friend Kip Thorne in 1985. Sagan was worried that his physics wasn't quite right. Like other science fiction writers, he was trying to get his heroine from earth to a distant location (in this case, the star Vega, and beyond) in the blink of an eye. Thorne suggested a wormhole, even though he had doubts as to whether a real wormhole (in contrast to Hawking's imaginary-time wormholes) could exist as a traversable tunnel in space-time. Afterwards, Thorne and his graduate students began pondering the reality of wormholes and initiated serious research on the topic.[2] Their fascination was partially due to the fact that wormholes, if they existed, could allow travel in time as well as in space. Under the right conditions, a wormhole might allow a traveler to return to his or her starting point before he or she left, and somehow change history.

The so-called grandfather paradox in theories of time travel is usually stated as follows: a time traveler could travel back in time and either kill his or her grandfather or otherwise prevent his or her grandparents from marrying before his or her parents were conceived, meaning that he or she would never have been born. But how, then, could he or she be there to prevent the marriage in the first place? A variation on this theme was the basic plotline of the movie *Back to the Future.* A trajectory in space-time that allows one to travel back in time is called a closed time-like curve. Such events constantly happen on the microscopic level, as an alternate but equivalent interpretation of the Casimir effect described in Chapter 5. Instead of considering a virtual particle-antiparticle pair being created and subsequently destroyed, one could think of a single particle moving forward and backward in time over and over again, like Bill Murray's character in the movie *Groundhog Day*, or *Star Trek: The Next Generation*'s Klingon, Worf, in the episode "Parallels." Could such time-like curves really be created at the macroscopic level, and if so, could they be exploited to permit travel by intelligent beings back in time?

Thorne and others found that the only way to make a traversable wormhole was to use exotic material with a negative energy density to hold the throat open. Negative energy seems like an oxymoron, an impossibility. Remember, though, that through the laws of quantum mechanics it is possible to "borrow something from nothing" in terms of energy. In theory, it is possible to create negative energy, although there is no technology known today which could generate it in the quantities required to make a usable wormhole. Thorne theorized, however, that it might be possible, in principle, for an extremely advanced technological society to be able to create negative energy in bulk. There was a serious catch—calculations by Thorne and postdoctoral student Sung-Won Kim showed that the moment the wormhole time machine was turned on, it would destroy itself in a tremendous explosion. They did find some cases in which the wormhole might survive the explosion, which left some hope of an advanced society creating a time machine.[3]

Hawking read Thorne's manuscript on the subject and vehemently disagreed. As Thorne explained, "There is little politeness in our community when one of us believes the other is wrong."[4] Hawking set out to prove why he thought Thorne was in error, resulting in the creation of the chronology protection conjecture. Hawking put forth a proof that "the laws of nature prevent the appearances of closed time-like curves."[5] In typical Hawking humor, he offered that there seems to be a "chronology protection agency, which prevents the appearance of closed time-like curves and so makes the universe safe for historians."[6] He later added, "the best evidence we have that time travel is not possible, and never will be, is that we have not been invaded by hordes of tourists from the future."[7] The paper was sent to Kip Thorne to review, who called it a tour de force. For Kip's sixtieth birthday, Hawking calculated the probability of success for a wormhole time machine, and came up with 1 part in 1060.[8]

RISING MEDIA STAR

In the spring of 1989 Hawking had been approached by LA producer Gordon Freedman about a possible film version of *A Brief History of Time.* Amblin Entertainment, Steven Spielberg's production company, was contacted for financing, which was granted based on the choice of the director, Errol Morris. Morris was an Academy Award-winning documentary creator, who had actually been a student of John Wheeler at Princeton and had a background in the history and philosophy of science.[9] Hawking's idea was for the film to be like his book, in other words, about the science, not the scientist. However, as shooting progressed and Hawking noted the number of interviews filmed with family and friends, he realized it would be half biography and said, "now I think that may have been a good thing."[10] Jane was contacted about appearing in the film, but understandably refused, as did Elaine Mason. Hawking himself would not discuss his personal life on camera.

The movie premiered in Hollywood on August 14, 1992, at the Academy of Motion Picture Arts and Sciences, following a cocktail party hosted by the ALS Association. Hawking thanked Errol Morris for making his mother into a movie star.[11] Although some reviewers criticized the glaring omission of his marital separation from the film, it was overall highly acclaimed. The film eventually garnered numerous awards, including the Grand Jury Prize and Filmmaker's Prize at the Sundance Film Festival, and the Filmmaker's Award from the National Society of Film Critics.

The following March the movie was again celebrated, in the form of the

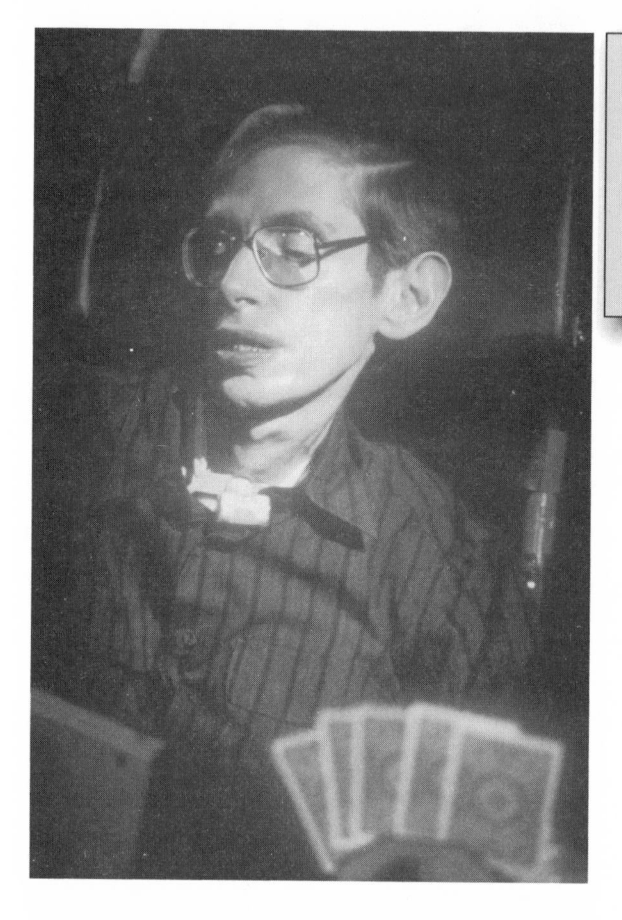

Stephen Hawking playing a holographic image of himself for an episode of Star Trek: The Next Generation, *1993. Photofest.*

home video version. At the release party, Hawking was introduced by none other than Leonard Nimoy, "Mr. Spock" from the original *Star Trek* series. After the party, Gordon Freedman explained to Nimoy that Hawking was a long-time fan of the show. In response, Nimoy contacted Rick Berman, the executive producer of *Star Trek: The Next Generation* and a cameo appearance was written for Hawking.[12] In an interesting twist, Hawking played a holographic version of himself in the episode "Descent," playing poker with holographic versions of Albert Einstein and Sir Isaac Newton, and the show's android, Data. Hawking bragged that he "beat them all, but unfortunately there was a red alert, so I never collected my winnings."[13] Hawking's presence on the set made a lasting impression on the crew and cast, with actor Brent Spiner, who portrayed Data, calling the event perhaps "the high point of my days on *Star Trek*."[14]

Hawking's very visible foray into the heart of pop culture led to more opportunities. His computerized voice was featured (with credit) in the song "Keep Talking" on Pink Floyd's *The Division Bell* album. A second, popular-level book of his was published by Bantam Books in 1993, a collection of essays, talks, and interviews, titled *Black Holes and Baby Universes and Other Essays.* Considered by many an easier book to read than its predecessor, one reviewer wrote, it "sprinkles his explanations with a wry sense of humor and a keen awareness that the sciences today delve not only into the far reaches of the cosmos, but into the inner philosophical world as well."[15]

With fame also came opportunities he could share with his colleagues. NEC, a Japanese information technology (IT) corporation, sponsored a lecture tour of Japan that included Hawking and string theorist David Gross. Along with their public lectures, Gross and Hawking were given a tour of the country, and, as Gross noted, "when you travel with Stephen you get to meet all sorts of people you would never meet otherwise—he opens all sorts of doors for you." Hawking was also adventurous in his travels, and on this particular trip insisted that they go to a karaoke bar. The group sang the Beatles' "Yellow Submarine," with Hawking joining in for the chorus with the phrase "Yellow Submarine." Gross later joked that Hawking probably still had a special "Yellow Submarine" button programmed.[16]

More serious work continued to occupy parts of Hawking's increasingly busy schedule. He edited two professional volumes in 1993, one a collection of his papers on black holes and the big bang. He introduced the volume by saying that

> with hindsight it might appear that there had been a grand and premeditated design to address the outstanding problems concerning the origin and evolution of the universe. But it was not really like that. I did not have a master plan; rather I followed my nose and did whatever looked interesting and possible at the time.[17]

The other work was a volume of technical articles on Euclidean quantum gravity, coedited by Gary Gibbons.[18] Of the thirty-seven influential and important papers contained in the book, sixteen were written or cowritten by Hawking, a testimony to his dominance in the field.

In 1994, Hawking and Roger Penrose held a debate about the nature of space-time and the prospects for a working theory of quantum gravity at the

Isaac Newton Institute for Mathematical Sciences at the University of Cambridge. Each gave three technical lectures presenting their philosophical and scientific points of view. Hawking took what he called the "positivist" viewpoint, explaining, "I don't demand that a theory correspond to reality because I don't know what it is. Reality is not a quality you can test with litmus paper. All I'm concerned with is that the theory should predict the results of measurements."[19] Such a philosophy is in keeping with his comments that imaginary time might actually be more important than what we perceive as real time. The lectures were published in 1996 in book form.

FAMILY CHANGES

Stephen and Jane finalized their divorce in the spring of 1995, and Jane spent some time in Seattle that July visiting Robert, who now worked for Microsoft. The press was soon filled with the news that Stephen and Elaine had set a wed-

Stephen and Elaine Hawking on their wedding day, 1995.
© Murray Andrew/Corbis Sygma.

ding date for September. Stephen had just become a member of the influential Aspen Center for Physics in Colorado, and at a special concert on July 8 at the Aspen Music Festival, he was asked to introduce the musical selections for the concert. He dedicated Wagner's *Siegfried Idyll* to his new fiancée. The press, once again sensing a scandal, brought immediate attention to the announcement. Some went so far as to question Elaine's motives for marrying Stephen, with crass insinuations about Stephen's amassed fortune (largely based on royalties from *A Brief History of Time*). Although he had made disparaging remarks about Stephen and Elaine after the break up of his own marriage, David Mason, Elaine's ex-husband and the originator of Hawking's wheelchair-mounted communication setup, came to Elaine's public defense.[20]

Upon returning home to the Cambridge home she now openly shared with Jonathan and Timothy, Jane received a letter from editor Susan Hill of Macmillan Publishers, asking about the possibility of her writing an autobiography. Jane had certainly pondered the possibility before. After Stephen had moved out, Jane put her energies into writing a book about her experiences renovating The Moulin, despite the fact that "various publishers were already clamouring" for her memoirs.[21] She found herself tricked into a deal by a devious agent, and rather than succumb to demands that she produce a tell-all, she waited until the terms of their contract ran out in the summer of 1994, then published her book, *At Home in France*, herself.[22]

The timing of Hill's offer, several days before Stephen's remarriage, seemed more appropriate to Jane. Now was certainly the time to find some closure on the past and move on, and perhaps sufficient time had passed to gain the calm insight that only comes with distance from pain. Stephen and Elaine exchanged vows in a private civil ceremony on September 16, 1995, followed by a church blessing, and Jane began writing her autobiography.

A Brief History of Time had been a runaway success, but there still remained the charges that it was not as accessible as a popular-level book should be. Hawking had an opportunity to revisit his seminal work, revising it and adding a new chapter on wormholes and time travel. It was also decided that it would benefit from copious, full-color illustrations, both photographs from the Hubble Space Telescope as well as computer-generated images that illustrated his abstract physical concepts. *The Illustrated A Brief History of Time* was released in November 1996.

Working for the rights of the disabled was also a priority for Hawking in the 1990s. In the summer of 1995, he gave a sold-out lecture at the 5,000-seat

Royal Albert Hall, with all the proceeds going to an ALS charity. He also lent the considerable power of his name and presence to "Speak to Me," an exhibit at the Science Museum, London, which displayed technology's benefits for the disabled.

Hawking began using the power of his fame to achieve scientific goals as well. In January 1997, he brought together a group of U.K. cosmologists from the Universities of Cardiff, Durham, Oxford, and Sussex, Imperial College, London, and the Royal Observatory in Edinburgh, and formed the COSMOS consortium. Led by Hawking, this group procured funding and expertise from Intel, Silicon Graphics Inc. (SGI), and British science organizations in order to create a state-of-the-art supercomputing facility for the study of cosmology. Hawking retains the title of Principal Investigator, and continues to broker contacts with SGI for improved technology.

The entire Hawking clan received surprising news in March. Lucy, who had been working in New York as a journalist, announced that she and her boyfriend, Alex Mackenzie Smith, a member of the United Nations Peace Corps in Bosnia, were expecting a child and would be living together in London. Jane and Jonathan were finally married on July 4, 1997, with their old friend and former pastor Bill Loveless presiding. Several months later, Lucy gave birth to her parents' first grandchild, William.

Stephen was busy giving birth to a project of his own. After the phenomenal success of *A Brief History of Time*, he had been approached by an old friend, David Filkin, who had been a member of his rowing team at Oxford. Filkin was the Head of Science and Features at BBC Television, and proposed making a several-part television series on Hawking's work. Stephen made it clear that he was tired of the sensational "brilliant brain trapped in a crippled body" programs that had been made in the past, and wanted a purely scientific work.[23] The BBC, in collaboration with American public television, produced six episodes. The series and a companion book, both titled *Stephen Hawking's Universe*, appeared in 1997. Hawking offered, "if we can unscrew the front panel of the universe and look behind, we might be able to figure out how the little wheels work and feel that we have some control over what is going on." In typical Hawking humor, he added, "Fortunately we aren't called upon to put the universe back together again."[24]

NATURE DOES WHAT IT ABHORS

That wry sense of humor was openly displayed at a public lecture in California, where he conceded yet another scientific wager. On September 24, 1991, he bet against John Preskill and Kip Thorne of Caltech concerning naked singularities. (Roger Penrose's theoretical prohibition against singularities not clothed in an event horizon was described in Chapter 4.) Although there was no definitive proof as to whether or not the cosmic censorship conjecture was true, Hawking was willing to bet it was, with two-to-one odds in his favor. The loser was to "reward the winner with clothing to cover the winner's nakedness." The clothing was "to be embroidered with a suitable concessionary message."[25] By 1997, theoretical calculations by Demetrios Christodoulou of Princeton, bolstered by computer simulations by Matthew Choptuik at the Center for Relativity at the University of Texas at Austin, suggested that under very specialized circumstances, a singularity might form without an event horizon.

However, the fine-tuning required to accomplish this made such an actual event very unlikely. Therefore, Hawking conceded on February 5, 1997, but with a twist. The clothing he presented to Thorne and Preskill was a T-shirt bearing the cartoon image of a naked woman loosely wrapped in a towel on which was written "Nature Abhors a Naked Singularity." Both Elaine Hawking and Carolee Thorne were aghast at this, but "Stephen has never been politically correct."[26] It was Hawking's way of demonstrating nature's reluctance to the apparent violation of the conjecture. However, Christodoulou revised his calculations, without previous approximations, and found that Hawking had probably conceded prematurely. Meanwhile, Hawking, believing he had lost the previous bet on a technicality, immediately made a new wager with Preskill and Thorne, with two caveats: the first, that any naked singularities must be created from generic rather than special conditions, and the second that the loser's message must be truly concessionary.[27]

In 1997, President Bill Clinton initiated the Millennium Evening series, consisting of eight lectures and cultural showcases spread over three years, hosted by the White House, and carried live over the Internet. The second event, held on March 6, 1998, featured a lecture by Stephen Hawking, titled "Imagination and Change: Science in the Next Millennium." He took the opportunity to warn of the impending dangers to society, such as overpopulation and unchecked electricity consumption. He predicted that "it is likely that we will be able to completely redesign" human DNA in the next millennium, and that although most people would probably approve a ban on human genetic engineering, he doubted such research could be prevented. As he explained, "Unless we have a totalitarian world order, someone will design improved humans somewhere." He also warned about the very real danger that the human race might destroy all life on earth:

> Even if we don't destroy ourselves completely there is the possibility that we might descend into a state of brutalism and barbarity like the opening scene of *Terminator*. But I'm an optimist. I think we have a good chance of avoiding both Armageddon and a new Dark Age.[28]

Once again, Hawking had drawn attention to the peril posed by nuclear weapons and the misuse of technology.

Ten years had passed since the release of *A Brief History of Time*, and in celebration, Bantam Books released an updated and expanded "Tenth Anniversary Edition." The decade had seen many changes in Hawking's life, some more positive than others. His fame had steadily grown over those ten years and he now seemed poised to take over the role of science's unofficial public spokesman from the late Carl Sagan. The public might hang on his every word, but some of his colleagues would have a more cynical view.

NOTES

1. Stephen Hawking, ed., *Stephen Hawking's A Brief History of Time: A Reader's Companion* (New York: Bantam Books, 1992), pp. 170–71.

2. Kip Thorne, *Black Holes and Time Warps* (New York: W. W. Norton and Co., 1994), pp. 483–90.

2. Kip Thorne, "Warping Spacetime," in *The Future of Theoretical Physics and Cosmology*, ed. G. W. Gibbons, E. P. S. Shellard, and S. J. Rankin (Cambridge: Cambridge University Press, 2003), pp. 101–102.

4. Ibid., p. 102.

5. S. W. Hawking, "Chronology Protection Conjecture," *Physical Review D* 46 (1992): 610.

6. Ibid.: 603.

7. Stephen Hawking, *Black Holes and Baby Universes and Other Essays* (New York: Bantam Books, 1993), p. 154.

8. Thorne, "Warping Spacetime," p. 103.

9. Gerald Peary, "Stephen Hawking: The Universe in a Mind," *New York* (September 1992): 126.

10. Kristine McKenna, "A Mind Over Matter: The Universe of Stephen Hawking," *Los Angeles Times* (August 19, 1992): sec. F.

11. Peary, "Stephen Hawking: The Universe in a Mind": 140.

12. Michael A. Lipton and Stanley Young, "Trek Stop," *People* (June 28, 1993): 81.

13. Stephen W. Hawking, *The Universe in a Nutshell* (New York: Bantam Books, 2001), p. 157.

14. Lipton and Young, "Trek Stop": 81.

15. Bill Sharp, review of *Black Holes and Baby Universes and Other Essays* by Stephen Hawking, *New York Times Book Review* (October 24, 1993): 22.

16. David Gross, "String Theory," in *The Future of Theoretical Physics and Cosmology*, ed. G. W. Gibbons, E. P. S. Shellard, and S.J. Rankin (Cambridge: Cambridge University Press, 2003), pp. 465–66.

17. Stephen Hawking, *Hawking on the Big Bang and Black Holes* (Singapore: World Scientific Publishing Co., 1993), p. 1.

18. G. W. Gibbons and S. W. Hawking, ed., *Euclidean Quantum Gravity* (Singapore: World Scientific, 1993).

19. Stephen Hawking and Roger Penrose, *The Nature of Space and Time* (Princeton, Princeton University Press, 1996), p. 120.

20. Richard Jerome, Vickie Bane, and Terry Smith, "Of a Mind to Marry," *People* (August 7, 1995): 45.

21. Jane Hawking, *Music to Move the Stars: A Life with Stephen Hawking* (London: Pan Books, 2000), p. 572.

22. Ibid., pp. 579–85.

23. David Filkin, *Stephen Hawking's Universe* (New York: Basic Books, 1997), p. 9.

24. Stephen Hawking, "Foreword," in Filkin, *Stephen Hawking's Universe*, p. xiii.

25. Thorne, "Warping Spacetime," p. 96.

26. Ibid., p. 98.

27. James Glanz, "New Proof Hides Cosmic Embarrassment," *Science* 276 (1997): 39.

28. Stephen Hawking "Remarks by Stephen Hawking," *White House Millennium Council 2000*, http://clinton4.nara.gov/Initiatives/Millennium/shawking. html (accessed October 23, 2001).

CHAPTER 11

PLAYS, P-BRANES, AND POLLS
Private Lives and Public Pronouncements

OPEN INFLATION, OPEN DEBATE

By the beginning of 1998, Hawking's reputation, among both his peers and the general public, was reaching new heights. This was widely apparent when a scientific paper coauthored with Cambridge colleague Neil Turok was accepted for publication in only three days.[1] Their collaboration had uncovered a new method of spawning an inflationary universe. In the classic inflation model, and in so-called new inflation, there was a very firm prediction about the universe—it was geometrically flat, meaning that the amount of matter in the universe was fine-tuned to a critical value, precariously perched between too much and too little. Just a dash more matter, and the universe would eventually collapse in on itself (i.e., it would be closed) whereas just a smidgen less and the universe would expand forever (i.e., it would be open).

Turok and colleagues Martin Bucher of Princeton and Alfred Goldhaber of SUNY Stony Brook had previously found a way to generate an open yet inflationary universe.[2] On the other hand, Hawking's earlier work with Jim Hartle on the no-boundary proposal had predicted a closed universe, which could not be reconciled with an inflationary model. Over tea, following a Cambridge seminar on open inflation, Turok and Hawking began talking about combining their models. They found that it was indeed possible to have the universe begin as a small, finite "pea" about one gram in mass, and then inflate into an infinite, open universe, such as current observations seemed to suggest. There was one serious catch to their model—it predicted a large number of possible universes, many of which were empty of matter. Hawking appealed to the anthropic principle once again, to weed out all such solutions that could not support the existence of intelligent life forms.

The paper drew immediate attention from the physics community, not all of it positive. Some criticized the duo for using the controversial no-boundary proposal while others were concerned about invoking the anthropic principle in a serious way. One of the paper's most instant and vocal critics was Hawking's old friend, Andrei Linde, now on the faculty of Stanford University in California. Shortly after Hawking and Turok released a preprint of their paper, Linde produced a long rebuttal of his own, followed soon thereafter by a reply from Hawking and Turok. Linde was seriously concerned that the model produced, at best, universes that were much too empty—they included about one-thirtieth the density of matter currently observed. Turok countered that their calculations were done with a simple model, and that a more realistic model should yield better results.[3]

The press caught wind of the dispute, and elevated it to the level of an intellectual war, instead of the more normal scientific give-and-take it was in reality. Articles appeared in *Science*, the *London Telegraph*, and *Manchester Guardian* in England, as well as the Stanford University online newsletter. Linde said of Hawking, "Stephen is an extremely talented person.... Sometimes, however—this is my interpretation—he trusts mathematics so much that he makes calculations first and interprets them later." He also compared Hawking's faith in mathematics to religion.[4] In another interview, he called Hawking "a very brilliant man," and pointed out that he has "come up with surprising conclusions, that at first, seem like they are wrong. But in several instances he turned out to be right. In other cases, he was wrong. We will just have to wait and see which it is this time."[5]

In April, Hawking traveled to Stanford at Linde's invitation and gave a seminar on the theory to an overflow crowd, and at the COSMO 98 Workshop on Particle Physics and the Early Universe in Monterey, California; in November, he took part in a debate with Linde and Alexander Vilenkin of Tufts University. Hawking justified his use of the anthropic principle, pointing out that "clearly, the universe we live in, didn't collapse early on, or become almost empty. So we have to take account of the anthropic principle, that if the universe hadn't been suitable for our existence, we wouldn't be asking why it is the way it is."[6] Hawking and Turok's paper is still cited in research papers as of 2004, so it would seem that their theory has yet to be dismissed.

THE PAST REVEALED

The end of the millennium was approaching, and in late 1998 the National Portrait Gallery of London planned an exhibit of one hundred photographs reflecting Britain in the twentieth century. Hawking was the only scientist among the ten influential persons (including musician David Bowie) selected to choose the photographs for the exhibit. He announced that his selections would focus on women and scientists, "the important members of society."[7] Among his eventual selections were photographs of Crick and Watson with a model of the DNA molecule, a portrait of physicist Paul Dirac, and an image of Prime Minister Margaret Thatcher.

Hawking's appearance on *Star Trek: The Next Generation* opened up other television appearances, doing voice-overs of himself for animated shows. He was featured in a 1999 episode of *The Simpsons*, "They Saved Lisa's Brain," in which he rescued Lisa from evil MENSA members. In a classic line, he told Homer that his theory of "a doughnut-shaped universe is interesting" and that he "may have to steal it." Hawking almost missed his cast call, as his wheelchair broke down two days before he was scheduled to fly from Monterey, where he was visiting, to Los Angeles. Chris Burgoyne, his graduate assistant, aided by a technician, worked a 36-hour shift in order to make the necessary repairs.[8]

In the episode "Infomercial" of the short-lived series *Dilbert*, a machine unwittingly created a black hole, and the character Dogbert kidnapped Hawking in order to repair space-time. In 2000, Hawking's cartoon image also appeared in *Futurama* in the episode "Anthology of Interest I," along with that of actress Nichelle Nichols of the original *Star Trek* series. Controversy also circled the use of Hawking's voice. While Pink Floyd had involved Hawking in the recording of their *The Division Bell* record, a computerized voice that people assumed was Hawking's turned up in a commercial for the Radiohead album *Amnesiac.* The same voice had been used on the earlier album *OK Computer*, and the band assumed fans would connect the commercial to their earlier album, not to Hawking.

Although his physical condition had been relatively stable for many years, Hawking underwent an operation in early 1999 to reroute his larynx so that there was less chance of food falling down the wrong pipe and into his lungs. This considerably reduced the risk of choking and allowed him greater enjoyment of meals. That same year he also enjoyed two more awards, the Naylor Prize and Lectureship in Applied Mathematics from the London Mathematical Society, and the Julius Edgar Lilienfeld Prize of the American Physical Society for a "most outstanding contribution to physics."

Despite the publicity machine that increasingly surrounded him, Stephen had made a point of trying to keep his private life just that. With the publication of Jane's autobiography, *Music to Move the Stars*, in August 1999, that was no longer possible. The book produced an expected furor, as intimate details of the Hawkings' marriage were made public. Book reviewers latched onto what they perceived as the sensational nature of a tell-all written by the ex-wife of "the world's most famous scientist." One article boldly exclaimed that Stephen had become the "victim of a bitter kiss-and-tell memoir by his former wife that has stunned the academic world."[9] Even the physics journal *Physics World* reviewed the volume, prefacing their review with a note that they were "not in the habit of reviewing books by non-physicists, but when the author was married to one of the most famous physicists of the twentieth century, we can make an exception."[10] Hawking did not publicly comment on the book, noting that he did not read biographies of himself.

Jane addressed the uproar in a postscript added to the paperback version of her book, published the following year. She sadly noted that the furor surrounding the book only affirmed her conviction that the false public persona of perfection that the family had maintained over the years had come to no good end. She herself was convinced that her two reasons for writing the book had been at least partially fulfilled. Her private aim had been to liberate herself from the past—it had done that, as well as remind her of how much she loved and cared for Stephen.[11] Her more public goal was to

> reveal the heart-breaking reality faced daily in an uncaring society at large by disabled people and their carers, the battles with officialdom, the lonely struggle to keep going and maintain a sense of dignity, the tiredness, the frustration, the anguished scream of despair.[12]

This goal also had been achieved in a small way, as Jane had been asked to address conferences on the subject of caring for the disabled.

The publicity surrounding Jane's book was an unmistakable indication of just how far Stephen's reputation had come, from brilliant physicist to true pop culture celebrity. One reporter noted how his iconic image of "pure, disembodied intellect...turned his every pronouncement into a front page story."[13] Hawking was even asked if he had donated his DNA for cloning, to which he replied in his usual humor, "I don't think anyone would want another copy of me."[14] It was, therefore, not completely unexpected that Hawking would become the subject of a play, *God and Stephen Hawking*, written by Robin Hawdon. Not only were Hawking and God characters in the play, but Jane Hawking, Newton, Einstein, the Queen, and the Pope were also portrayed, most by the same actor. Hawking was sent the script to review in early 1999, and found it ridiculous and rather embarrassing. His response was to ignore it in the hope that it would never reach production. However, when the author added details from Jane's book, Hawking called it "deeply offensive and an invasion of my privacy," but decided not to pursue legal action because he felt it would attract undue attention to "a stupid and worthless play."[15] The play received mixed reviews when it opened in August 2000, but Lucy Hawking later noted that when she saw her family life portrayed on stage she was both "horrified and mesmerized," describing her "insane urge to climb on the stage and join them."[16]

MILESTONES ACHIEVED

Despite the negative effects that fame had on his desire to protect his private life, Hawking did not withdraw from the public arena. In December 1999 he taped an interview with CNN's Larry King, which was broadcast on Christmas. When asked how he was planning to spend New Year's Eve, he replied that they were going to have a *Simpsons*-character costume party. The best part, for him, was that he could go as himself.[17] The following May, he openly contradicted Prince Charles' public statements against genetically modified (GM) food. The Prince was concerned that "tampering with nature could cause great harm to the world," while Hawking countered that, "I don't think you can outlaw research and development because it can be put to good use." He added that in fifty years people "will wonder what all the fuss about GM food was all about."[18] Hawking even provided a televised tribute for presidential candidate Al Gore at the Democratic National Convention in August 2000.

Hawking's reputation was benefiting the University of Cambridge in palpable ways. He had become a funding magnet for the university and especially the Department of Applied Mathematics and Theoretical Physics. By 2000, it had received a brand new, $177 million building, with a state-of-the-art, wheelchair-accessible office for Hawking.[19] Intel Corporation cofounder Gordon Moore and his wife, Betty, donated $12.5 million for the creation of a physical sciences and technology library, which was touted as the new home of the Hawking Archive of personal papers, including an early draft of *A Brief History of Time.* Hawking's connection to Intel was through the software they had donated for his wheelchair-mounted computer communication system.[20]

It had become customary for colleagues and students of physicists to celebrate their colleagues' sixtieth birthdays with a special seminar or conference and volume of papers celebrating the sum of their academic work. The tradition was known as a *festschrift*, a German term meaning "feast writing." On June 3, 2000, Hawking gave a lecture at Caltech for the KipFest Saturday Science Talks in honor of long-time friend Kip Thorne's sixtieth birthday, and in March 2001 he took part in a similar celebration for University of California, Santa Barbara, string theorist and 2004 Physics Nobel Laureate David Gross. Hawking himself was still shy of his sixtieth birthday, but despite the dire prediction of his physicians when first diagnosed with ALS, it seemed he might indeed live long enough to enjoy a festschrift of his own.

Not only was Hawking keeping to a professional conference schedule, but he also coupled it with a worldwide public speaking circuit. His lectures drew tremendous crowds, unheard of for a theoretical physicist since Einstein. In South Korea he entertained a crowd of 4,000 at Seoul National University in September 2000, and in April 2001 a talk in Granada, Spain, was projected onto screens in a science park.[21] In October 2000 Hawking received a Sarojini Damodaran International Fellowship to cover travel expenses to India. Not only did he attend the Strings 2001 conference in January 2001 in Mumbai, but afterwards he also traveled to Delhi to deliver the Albert Einstein lecture to a standing-room-only crowd of nearly 4,000, including many local dignitaries.

Hawking continued to be a symbol of success for the handicapped, and used his celebrity for the benefit of others. In 1999 he joined eleven other international dignitaries, including South African Archbishop Desmond Tutu, in signing the "Charter for the Third Millennium on Disability." The World Assembly of Rehabilitation International adopted the document, which called for "governments to demonstrate the political will to prevent easily avoidable conditions and illnesses which cause disability," including meningitis and landmines.[22] In February 2001, he gave a public lecture in Lady Mitchell Hall, Cambridge, which raised thousands of dollars for the Newnham Croft Primary School's "Raise the Roof" campaign to fund a new extension. He raised public awareness of the technology available for the handicapped by proudly advertising the new Quantum Jazzy 1400 wheelchair provided to him by Pride Mobility. He compared the new chair to a Ferrari, and added that it "will also keep my nurses fit as they try to keep up." His personalized license plate reflected his favorite drink: T4SWH (tea for Stephen W. Hawking).[23] The public's appetite for information on Hawking was temporarily sated by an hour-long documentary, *The Real Stephen Hawking*, which aired on BBC 4 in the summer of 2001.

Timothy Hawking was now a college student at the University of Exeter, following in his mother's footsteps as he studied French and Spanish. He still enjoyed spending time with his father, and opened Stephen's horizons to include Formula One racing and a Depeche Mode concert. Of the latter, Stephen said, "I really enjoyed it even though I was sitting in front of the speakers, and my ears were ringing for the next twenty-four hours."[24] Although his musical taste remained largely operatic, Hawking had attended a number of rock concerts over the years, including Pink Floyd and Tracy Chapman. In December 1990 he was attending a conference in Brighton, and his students used his name (and celebrity) to procure a handful of free tickets to a sold-out Status Quo concert being held next door. Unfortunately, Hawking himself found the concert terrible and left after only twenty minutes.[25] Lucy also continued to enjoy her relationship with her father, including his uncanny ability to buy "gifts of beautiful clothes which invariably fit perfectly." She said that it "means more to me that he knows what size I am, and not just what size galaxies are."[26]

ALL STRUNG OUT

The trademark Hawking adventurism continued in full bloom, but another of his well-known personality traits was showing signs of softening, at least in the scientific arena. Theorist and friend Leonard Susskind has referred to him as "by far the most stubborn and infuriating person in the universe."[27] This quality of the Hawking persona has been nowhere more visible than in his public disparaging of string theory. Originally developed in the late 1960s and 1970s as an alternative to the standard model of particle physics, string theory visualizes elementary particles, such as quarks and electrons, as unique frequencies of vibration of tiny subatomic objects called strings. In this way, the different particles are unified as diverse modes of a single fundamental object. String theory languished in a sea of mathematical problems and an embarrassing lack of predictive power until 1984, when Michael Green of Queen Mary College in London and John Schwarz of Caltech combined string theory with that of supersymmetry, creating superstrings. The downside for many scientists was that superstrings were mathematically consistent in ten dimensions, while the real universe seemed to exist in only four (three of space and one of time). No problem, the string theorists claimed, because the extra dimensions were curled up, or "compactified," into tiny little knots far too small to be observed. As strange as this idea might seem, string theory was not the first time extra dimensions were theorized in an attempt to achieve unification in particle physics. In the 1920s, physicists Theodore Kaluza and Oskar Klein had invoked an extra, hidden dimension of space in order to explain electromagnetism as arising from the geometry of space-time—an early attempt to unify electromagnetism and gravity.

Superstrings soon became the new fashion in theoretical physics, pushing aside supergravity in its wake. Some heralded it as the best chance for a theory of everything, while on the other hand it also seemed "a frustrating collection of folklore, rules of thumb, and intuition." It was also ironic that this supposedly unifying theory itself appeared so disunited.[28] There were at least five different string theories, with vastly different properties. In addition, there were still no experimentally verifiable predictions from superstrings. Thus, the physics community tended to be polarized on the subject, with "some closed-mindedness on one side and a certain amount of hubris on the other."[29] Hawking was certainly not shy about his opinion of string theory. In his 1994 debate with Roger Penrose, he flatly stated that "string theory has been oversold" and that "reports of the death of supergravity are exaggerations."[30] He further accused that string theory "has been pretty pathetic: string theory cannot even describe the structure of the Sun, let alone black holes."[31] While string theorists weren't quite as pessimistic, they had to acknowledge that there were, indeed, serious questions that remained in the early 1990s. New life was suddenly breathed into string theory, from a most unlikely source.

Leonard Susskind and other string theorists were deeply troubled by

Hawking's earlier prediction that information was forever lost in a black hole, because it seemed to say that there were serious questions with quantum mechanics, and string theory was rooted in the conventional rules of quantum mechanics. Theorists found that they could model certain black holes as very excited strings, and that the length of the string was related to the surface area of the black hole's horizon. But Hawking (and Bekenstein) had earlier demonstrated a relationship between the surface area of a black hole's horizon and its entropy. In 1996, Andrew Strominger and Cumrun Vafa at Harvard were able to derive the exact entropy formula for stringy black holes, giving string theorists hope that the information-loss paradox might be solvable through string theory.[32] Would Hawking be convinced? Shortly after Strominger and Vafa published their results, Joe Polchinski, one of the organizers of the Strings '96 conference, received an email with Hawking's return address. The email gave the title for Hawking's upcoming talk at the conference as "I was wrong all along. String theory is right and information is not lost down black holes." Several days later, the imposter confessed, and the real Hawking sent an email with the talk's true title: "Why I Have Not Changed My Mind."[33]

Regardless of Hawking's continued skepticism, string theorists pressed on, largely due to a groundbreaking discovery by Edward Witten of Princeton. He suggested that the five different string theories were actually special cases of a much more fundamental, underlying theory in eleven dimensions.

In addition, supergravity was also included under the umbrella of this new M-theory. A definitive reason for the name has never been given, with Witten suggesting "magic, mystery, or membrane."[34] Others have suggested it stands for "mother of all theories." At the heart of M-theory are what physicists call dualities, connections between theories which themselves appear different but nevertheless seem to lead to the same physical results. M-theory also predicts that besides point-like (zero-dimensional) particles and one-dimensional strings, there are two-dimensional membranes. There are also higher dimensional p-branes, (or simply branes for short), where p stands for the number of dimensions of space.

Many of the calculations involving branes and the M-theory dualities were most tractable in a special model of space-time called anti-de Sitter.[35] Part of the reason for this is that supersymmetry, the heart of the theory of quantum gravity known as supergravity, is unbroken in anti-de Sitter space-time.[36] However, anti-de Sitter space-time is not a realistic model for our universe. For example, a seminal paper[37] by Hawking and Don Page published in

1983 on the thermodynamic properties of anti-de Sitter space-time found serious instabilities in such geometries. If p-branes could be studied in the more realistic de Sitter space-times (such as those utilized in inflationary models), it would be a great step forward. However, since de Sitter space-time breaks supersymmetry, there is not yet a consistent quantum gravity theory for this geometry. A natural starting point has been Gibbons and Hawking's aforementioned infiuential 1977 paper on the thermodynamic properties of the de Sitter cosmological horizon. Hawking himself became involved in studying quantum gravity in de Sitter space, including five papers published with student Raphael Bousso.[38]

BRANE NEW WORLD

The idea of this mysterious, over-arching M-theory was enthusiastically embraced by many researchers. Even Hawking felt there was something to this new idea. He believed that not taking "this web of dualities as a sign we are on the right track would be a bit like believing that God put fossils into the rocks in order to mislead Darwin about the evolution of life."[39] Despite its promise, M-theory was admittedly incomplete. Hawking described it as a jigsaw puzzle, in that it is "easy to identify and fit together the pieces around the edges but we don't have much idea of what happens in the middle,"[40] where the edges of the puzzle correspond to the various string theories and supergravity. Regardless of these ongoing issues, researchers were able to successfully apply branes to certain particular models, including black holes. For example, black holes could be pictured as branes intersecting. A particle falling into such a brane black hole appears as a closed loop of a string striking one of the branes. The result is excited waves in the brane. Conversely, a piece of an excited brane can break off as another closed loop of string, corresponding to a particle emitted by the black hole. Most interesting is the suggestion that these waves on the brane would store information about the black hole's formation, appearing to avoid the dreaded information-loss paradox. But while string and brane theorists talked among themselves about a solution to the problem, Hawking himself was not publicly convinced.

Models of brane worlds—four-dimensional branes living in a higher dimensional space-time called the "bulk"—became more common in the scientific literature. They differed in the number and size of the extra dimensions (whether they were compactified into tiny knots, like in string theory, or infinitely large). One of the most important models was developed by Lisa Randall of the Massachusetts Institute of Technology (MIT) and Princeton and Raman Sundrum of Boston University in 1999,[41] and was used in an attempt to tackle one of the fundamental lingering questions in the path of unifying the forces, namely why gravity is so much weaker than the other forces. Most people have the mistaken impression that gravity is strong, especially when trying to get out of bed in the morning, or climb several flights of stairs. But the fact that both actions are done so routinely, clearly shows the weakness of gravity. The simple action of hanging a note on the refrigerator using a magnet also demonstrates gravity's inherently wimpy nature. The electromagnetic force that holds our atoms together is 10^{35} times more powerful than gravity. This relative weakness of gravity is called the hierarchy problem.

Higher dimensional theories to describe gravity were used with extreme caution in the past, because it was thought that if space had more than three dimensions, gravity would significantly differ in its behavior from what we measure in the lab. In other words, rather than drop as the inverse square of distance (twice the distance giving one fourth the strength), it might drop as the inverse cube of distance (twice the distance giving one eighth the strength), or some other relationship easily ruled out by experiment. Randall and Sundrum discovered that a particular model with a four-dimensional brane living in an infinite fifth dimension seemed to solve the hierarchy problem without changing the behavior of gravity we observe. They found that while matter, energy, and three of the forces would be stuck on the brane, gravity would leak off just enough to explain its weak appearance, yet not so drastically that current experiments would notice the difference in gravity's behavior.[42]

Between 1998 and 2002, Hawking and his current and former graduate students authored a number of influential papers melding the no- boundary proposal and brane theory (especially the Randall-Sundrum model) to derive new brane black holes and brane-based inflation models.[43] However, despite his newfound appreciation of M-theory, Hawking warned that string theorists should not think that their belief in strings as the ultimate answer had been verified. In his view, the dualities suggested that "string theory, p-branes, and supergravity are all on a similar footing. None of them is the whole picture, but each are valid in different, but overlapping regions." He also suggested that while it may be possible that M-theory (when and if its complete properties are discovered) might produce a true fundamental theory, it was also possible that theoretical physics is like a patchwork quilt, with "apparently different theories that are valid in different regions, but that agree on the overlaps."[44]

While Hawking's research was soaring into higher dimensions and inflating brane worlds, his public statements were about to become grounded in heated debate. In the fall of 2001, he issued a series of opinions during interviews on matters related to science and society. He told the German magazine *Focus* that humans needed to modify their DNA in order to keep up with computers. Otherwise, there would be the real danger that intelligent machines "will develop and take over the world."[45] The threat posed by biological and nuclear warfare was so serious, Hawking warned *The Daily Telegraph*, that humanity must create space colonies in order to insure its own survival. In his words, "although September 11 was horrible, it didn't threaten the survival of the human race.... The danger is that, either by accident or design, we create a virus that destroys us."[46] Although clearly outside his area of expertise, Hawking's opinions were considered newsworthy because he was perceived by some as representing science, and scientists, in general. The public was "seduced by his achievement into believing that whatever he has to say on any topic must be worth listening to—even as Newton was pressed into public service as Master of the Mint and Einstein was sought out to be Israel's head of state."[47]

Not everyone shared the belief that Hawking's opinion mattered. *Physics World* polled a series of physicists as to whom they thought was the top physicist of all time. Einstein received 119 votes, in stark contrast to Hawking's single vote.[48] In a larger survey by *PhysicsWeb*, Isaac Newton edged out Einstein, but Hawking still trailed, at sixteenth place.[49] Hawking's pronouncements were openly challenged by Dr. Benny Peiser, senior lecturer in Social Anthropology at John Moores University, Liverpool. As an author who has written extensively on the influence of asteroid impacts on human evolution, he criticized Hawking's "predictions of terrestrial doom," pointing out that in his opinion they were becoming "increasingly wide-ranging and unreasonable," referring to Hawking's publicized opinion of a year prior that the greenhouse effect threatened to turn Earth into another boiling-hot Venus.[50] Sue Mayer, director of Genewatch, a policy research group, complained that Hawking was "trying to take the debate about genetic engineering in the wrong direction.... It is naive to think that genetic engineering will help us stay ahead of computers."[51] Peter Coles, an astronomy professor at the University of Nottingham, complained that

> coffee-time talks in physics departments often come up with the same topic: it's very difficult to get anybody to say anything critical of him. But to have somebody like that in an establishment that runs on peer review isn't healthy. The trouble is, people fear that they will be thought of as jealous.[52]

Some countered that there was an unjust backlash occurring against Hawking simply because of his fame. Regardless of the reason, the intense media attention focused on Hawking was well-timed, as he prepared to describe for a general public his newfound appreciation for higher-dimensional theories and the "universe in a nutshell."

NOTES

1. S. W. Hawking and N. Turok, "Open Inflation Without False Vacua," *Physics Letters* B425 (1998): 25–32.
2. M. Bucher, A. S. Goldhaber. and N. Turok, "Open Universe From Inflation," *Physical Review D* 52 (1995): 3314–37.
3. Andrew Watson, "Inflation Confronts an Open Universe," *Science* 279 (1998): 1455.
4. Tom Yulsman, "Give Peas a Chance," *Astronomy* (September 1999): 38–39.
5. David F. Salisbury, "Hawking, Linde Spar Over Birth of the Universe," *Stanford Report Online*, April 29, 1998, http://news-service.stanford.edu/news/1998/april29/hawking.html (accessed July 28, 2004).
6. S. W. Hawking, "A Debate on Open Inflation," in *COSMO-98: Second International Workshop on Particle Physics and the Early Universe*, ed. David O. Caldwell (College Park, MD: American Institute of Physics, 1999), p. 21.
7. *PhysicsWeb*, "Hawking and Rotblat Choose Their Favourites," November 13, 1998, http://physicsweb.org/article/news/2/11/5/1 (accessed August 1, 2004).
8. Carol Kennedy, "Making More of Time," *Guardian Unlimited*, August 23, 1999, http://www.guardian.co.uk/Archive/Article/0,4273,3894899.html (accessed August 7, 2004).
9. Nicole Veash, "Ex-wife's Kiss-and-tell Paints Hawking as Tyrant," *India Express*, August 3, 1999, http://www.expressindia.com/ie/daily/19990803/ile03011.html (accessed March 18, 2003).
10. Matin Durrani, "Emerging from Hawking's Shadow," *PhysicsWeb*, November 1999, http://physicsweb.org/articles/review/12/11/1/1 (accessed March 27, 2003).
11. Jane Hawking, *Music to Move the Stars: A Life with Stephen Hawking* (London: Pan Books, 2000), p. 593.

12. Ibid., p. 2.

13. Robin McKie, "Master of the Universe," *The Observer*, October 21, 2001, http://observer.guardian.co.uk/comment/story/0,6903,577783,00.html (accessed October 23, 2001).

14. Peter Laufer, "Stephen Hawking Visits Silicon Graphics," *Silicon Graphics*, May 1998, http://www.sgi.com/features/1998/may/hawking (accessed October 22, 2001).

15. Matin Durrani, "Hawking Slams 'Stupid worthless' Play," *Physics World* (August 2000): 8.

16. Elizabeth Grice, "Dad's Important, But We Matter, Too," *The Telegraph*, April 13, 2004, http://www.telegraph.co.uk/arts/main.jhtml?xml=/arts/2004/04/13/bohawk13.xml (accessed May 25, 2004).

17. Stephen Hawking, interviewed by Larry King, *Larry King Live Weekend*, Cable News Network, December 25, 1999.

18. *BBC News*, "Hawking Rejects Prince's Science Concerns," May 18, 2000, http://news.bbc.co.uk/1/hi/uk/753326.stm (accessed October 23, 2001).

19. Jan Moir, "Meet Mr. Universe," *The Daily Telegraph*, October 20, 2001, http://www.smh.com/au/news/0110/20/review/review9.html (accessed October 22, 2001).

20. Toni Feder, "Cambridge to Get New Library, Hawking Archive," *Physics Today* (December 1988): 51.

21. Stephen Hawking, "Information Archive," *Professor Stephen Hawking's Web Pages*, http://www.hawking.org.uk/info/news2001.html (accessed May 25, 2004).

22. *BBC News*, "Health Call for Global Disability Campaign," September 8, 1999, http://news.bbc.co.uk/1/hi/health/441912.stm (accessed October 23, 2001).

23. Terri Rozaieski, "Visiting with Stephen Hawking," *Pride Mobility*, http://www.pridemobility.com/pridewebtalk/Stephen_Hawking/stephen_hawking.html (accessed November 9, 2001).

24. Stephen Hawking, "Questions and Answers," *Professor Stephen Hawking's Web Pages*, http://www.hawking.org.uk/about/qa.html (accessed September 14, 2004).

25. Recoil, "Celebrity Squares—Professor Stephen Hawking," *Shunt—the Official Recoil Website*, http://www.recoil.co.uk/report/edit/squares/hawkingfull.htm (accessed September 14, 2004).

26. McKie, "Master of the Universe."

27. Leonard Susskind, "Twenty Years of Debate With Stephen," in *The Future of Theoretical Physics and Cosmology*, ed. G. W. Gibbons, E. P. S. Shellard, and S. J. Rankin (Cambridge: Cambridge University Press, 2003), p. 330.

28. Michio Kaku, *Introduction to Superstrings and M-Theory*, 2nd ed. (New York: Springer-Verlag, 1999), p. 5.

29. Nick Warner, "Gauged Supergravity and Holographic Field Theory," in *The Future of Theoretical Physics and Cosmology*, ed. G. W. Gibbons, E. P. S. Shellard, and S. J. Rankin (Cambridge: Cambridge University Press, 2003), p. 511.

30. Stephen Hawking and Roger Penrose, *The Nature of Space and Time* (Princeton: Princeton University Press, 1996), p. 4.

31. Ibid., p. 123.

32. A. Strominger and C. Vafa, "Microscopic Origin of the Bekenstein-Hawking Entropy," *Physics Letters* B379 (1996): 99–104.

33. Joe Polchinksi, "M Theory and Black Hole Quantum Mechanics," in *The Future of Theoretical Physics and Cosmology*, ed. G. W. Gibbons, E. P. S. Shellard, and S. J. Rankin (Cambridge: Cambridge University Press, 2003), pp. 302–303.

34. Robert Naeye, "Delving into Extra Dimensions," *Sky and Telescope* (June 2003): 38–44.

35. The properties of anti-de Sitter space are discussed in Appendices A and D.

36. See Appendix C for details on symmetry breaking.

37. S. W. Hawking and D. N. Page, "Thermodynamics of Black Holes in Anti-de Sitter Space," *Communications in Mathematical Physics* 87, no. 4 (1983): 577–88.

38. For a description of their collaboration, see Raphael Bousso, "Adventures in de Sitter Space," in *The Future of Theoretical Physics and Cosmology*, ed. G. W. Gibbons, E. P. S. Shellard, and S. J. Rankin (Cambridge: Cambridge University Press, 2003).

39. Stephen Hawking, *The Universe in a Nutshell* (New York: Bantam Books, 2001), p. 57.

40. Ibid., p. 174.

41. Lisa Randall and Raman Sundrum, "An Alternative to Compactification," *Physical Review Letters* 83, (1999): 4690–93.

42. Their theory was based on a special duality used in brane theory, known as the AdS/CFT duality, and the extra "bulk" dimension was of a particular variety called anti-de Sitter space. Both are explained in Appendix D.

43. The two most widely cited are A. Chamblin, S. W. Hawking and H. S. Reall, "Brane-World Black Holes," *Physical Review D* 61 (2000): 065007, and S. W. Hawking, T. Hertog, and H. S. Reall, "Brane New World," *Physical Review D* 62 (2000): 043501.

44. Stephen W. Hawking, "Is Information Lost in Black Holes?" in *Black Holes and Relativistic Stars*, ed. Robert Wald (Chicago: University of Chicago Press, 1998), pp. 231–32.

45. Nick Paton Walsh, "Alter Our DNA or Robots Will Take Over, Warns Hawking," *The Observer*, September 2, 2001, http://www.observer.co.uk/uk_news/story/0,6903,545653,00.html (accessed October 23, 2001).

46. *Guardian Unlimited*, "Space Colonies Needed For Human Survival," October 16, 2001, http://www.guardian.co.uk/Archive/Article/0,4273,4278507,00.html (accessed November 10, 2001).

47. Chet Raymo, "Stephen Hawking and the Mind of God," *Commonweal* (April 6, 1990): 218.

48. Matin Durrani and Peter Rodgers, "Physics: Past, Present, Future," *Physics World* (December 1999): 7–13.

49. *PhysicsWeb*, "Newton tops PhysicsWeb Poll," November 16, 1999, http://physicsweb.org/articles/news/3/11/16/1 (accessed May 25, 2004).

50. Robert Roy Britt, "Stephen Hawking's Doomsday Predictions Called Regrettable Hype," Space.com, October 16, 2001, http://www.space.com/news/hawking_rebuttal_011016.html (accessed October 23, 2001).

51. Walsh, "Alter Our DNA or Robots Take Over, Warns Hawking."

52. Charles Arthur, "The Crazy World of Stephen Hawking," *The Independent*, October 12, 2001, http://www.jodkowski.pl/nr/Independent001.html (accessed March 18, 2003).

CHAPTER 12

BOOKS AND BETS

The Universe in a Nutshell and the End of a Paradox

IN A NUTSHELL

After the success of *A Brief History of Time*, Hawking was bombarded with requests for a sequel. He had declined to write what might be seen as a "Son of Brief History" or "A Slightly Longer History of Time" because he wanted to devote time to his research. However, with the new millennium, he began to realize that "there is room for a different kind of book that might be easier to understand."[1] He therefore set to work on *The Universe in a Nutshell*, which outlined his research since *A Brief History of Time* for the general public, including his research on brane worlds. Learning from the experiences of people who claimed they got stuck in the first few chapters of his previous book, he decided the new work would not have a linear layout, but would have separate, independent topics, which could be read in any order (or not at all), after mastering a relatively small core of essential material.[2] Along with topics from theoretical physics, Hawking also included his opinions on those controversial issues that had garnered him mixed press during the past few years, including the future of humanity and artificial intelligence. The book was launched in Munich, Germany, in October 2001 with a public lecture, followed by the British debut the following month.

The title of the book was taken from Shakespeare's *Hamlet*: "I could be bounded in a nutshell and count myself a king of infinite space." Hawking had grasped onto the analogy of a nutshell in an attempt to better explain the no-boundary proposal, which had confounded readers of *A Brief History of Time*. In imaginary time, the history of the universe would be pictured as roughly spherical, but it could not be perfectly round, as that would lead to permanent inflation, rather than just a brief episode of exponential expansion. In addition, the surface of the "nut" would have tiny ripples, corresponding to the slight deviations in temperature of the cosmic background radiation, evidence of the "seeds" which led to the formation of structures such as galaxies in the early universe. Therefore, in the Euclidean (imaginary time) picture, the history of the universe looked rather like a walnut shell. Physicist Joseph Silk noted in his review of the book, "I have still not completely got to grips with imaginary time, but the nutshell metaphor conjures up a much warmer image."[3]

Hawking extended the idea to brane worlds, the difference being that the inside of the nut would be filled with the bulk, the higher fifth dimension, with other dimensions (if they exist) curled up into an extremely tiny knot. Quantum fluctuations would allow brane worlds to spontaneously appear, like "bubbles of steam in boiling water." Most of the small bubbles would simply collapse, but some might expand large enough to become universes like our own. If we lived on the surface of such a brane, we would think that the universe was expanding on the surface of this bubble. As Hawking wryly put it, "Let's hope there's no one with a cosmic pin to deflate the bubble."[4] He also provided his personal philosophy on brane theory, stating that the question "'Do extra dimensions really exist?' has no meaning. All one can ask is whether mathematical models with extra dimensions provide a good description of the universe."[5]

Reviews for the book were generally favorable, and while it was not the runaway success of its predecessor, *The Universe in a Nutshell* was a bestseller. In June 2002, Hawking was awarded the Aventis Book Prize for excellence in popular science writing. Psychiatrist Raj Persaud, the Chair of the Judging Panel, complimented Hawking for making "a real effort to enliven the subject through readable text and clear illustration," and added that even readers who do not comprehend the entire book "will still gain so much by a story told by an extraordinary mind."[6] Hawking stated that he hadn't expected to win the prize, "after all, my previous book didn't win any prizes, despite selling millions. But I am very pleased to have had better luck this time." He added that "science writing really can have an impact on how we live. Wherever I go all around the world, people want to know more. This has helped raise the profile of science."[7]

AN UNEXPECTED BIRTHDAY

Several months after *The Universe in a Nutshell* was released, Hawking celebrated his much-anticipated, yet seemingly miraculous, sixtieth birthday. Brian Dickie of the Motor Neuron Disease Association remarked, "to live as long as he has is just remarkable. He is the British record holder for survival. He is a wonderful example to everyone with the disease."[8] But Hawking had done far more than merely survive ALS for nearly 40 years; he had led a remarkably productive life, rivaling or even exceeding that of able-bodied colleagues. His friends and colleagues held a festschrift for him in Cambridge, beginning on January 7, 2002, featuring a day of popular-level lectures (including one by Hawking himself, entitled "Sixty Years in a Nutshell," in which he requested that the formula for black hole entropy be eventually placed on his tombstone), and several days of scientific papers. Gary Gibbons, Hawking's friend, colleague, and one of the organizers, explained that the event was meant to "look back on the immense contribution that Stephen has made to many areas of gravitational physics and cosmology."[9] Long-time friend and colleague Roger Penrose noted that Hawking had "officially

Stephen Hawking at the Cambridge symposium held in honor of his 60th birthday, 2002. Getty Images.

become an old man" and was now better able to "get away with saying such outrageous things." However, Penrose noted that Hawking "has always done that kind of thing, but he can perhaps feel a little bolder in this even than before."[10] The event (and the resulting volume of papers) was entitled *The Future of Theoretical Physics and Cosmology*, a twist on the title of his Lucasian Inaugural lecture, "Is the End in Sight for Theoretical Physics?" The popular-level lectures were later broadcast as *The Hawking Lectures* on BBC.

Besides the scientific celebration, there was also a rather large birthday party, attended by a crowd of two hundred, including most of Hawking's twenty-four current and previous graduate students. The festivities featured a Marilyn Monroe impersonator serenading Hawking with "I Want to Be Loved By You," and a choir that included ex-wife Jane, former and current students, and U2 guitarist The Edge, and was conducted by Jane's second husband, Jonathan. Jane was especially happy to be involved, and said, "because I still think that, with the exception of our children, the greatest achievement of my life was helping keep him alive."[11] This joyous celebration almost didn't happen, because shortly after Christmas, Hawking "had an argument with a wall," and not unexpectedly, "the wall won."[12] He was traveling on a cobbled street near his home, lost control, and crashed his wheelchair into a wall. His broken hip had to be operated on with only an epidural anesthetic (rather than a general anesthetic) because of his medical condition. He explained to friends that it was "like hearing a Black and Decker drill."[13]

During his hospitalization in 1985 when he received his tracheotomy and lost his natural voice forever, he had dreamed of flying in a hot air balloon. Two months after his sixtieth birthday, Elaine made his dream a reality. Hawking and his wife took a thirty-minute flight in a specially designed balloon basket. Shortly afterward, Hawking took a more traditional mode of flight to America, for a two month visit to California and Oregon. While there, he gave both public and technical talks, as had become his standard procedure. After returning to Cambridge and taking part in Martin Rees' festschrift celebration, he journeyed to China in August for more dual-level presentations.

The phenomenal success of Hawking's popular-level books and lectures left him open to what he considered unscrupulous publishing practices. He filed a complaint with the U.S. Federal Trade Commission in order to stop New Millennium Press from publishing a 1989 series of Cambridge lectures and material from *A Brief History of Time* as a new book—*The Theory of Everything*. He claimed that passing off old material as a new book was "a fraud on the public."[14] Although Hawking's complaint was ultimately unsuccessful in preventing the book's publication, or that of *The Illustrated Theory of Everything* in 2003, his official Web site still proclaims that he "has not endorsed this book" and that he "would urge you not to purchase this book in the belief that Professor Hawking was involved in its creation."[15] Hawking did publish a book of his own making in 2002, *On the Shoulders of Giants*,[16] a thick compilation of excerpts from the great works of Copernicus, Galileo, Kepler, Newton, and Einstein, along with biographical information and commentary.

"I suggested we might find a complete unified theory by the end of the century," Hawking reminisced for a reporter in April 2002. "OK, I was wrong," he laughingly added. He then updated his prediction with "I still think there's a 50–50 chance that we will find a complete unified theory in the next twenty years."[17] But Hawking was waxing much more philosophical several months later in a talk for the Paul Dirac Centennial Celebration at Cambridge. As Dirac, a Nobel Prize winner in physics, had greatly contributed to the understanding of both relativity and quantum mechanics, Hawking was a natural choice for a speaker. However, what he said about the possibility of developing a "unified theory" surprised some:

> M-theory has made me wonder if this is true. Maybe it is not possible to formulate the theory of the universe in a finite number of statements.... Some people will be very disappointed if there is not an ultimate theory that can be formulated as a finite number of principles. I used to belong to that camp, but I have changed my mind.[18]

As his friend Jim Hartle had once said, "if it's short enough to be discoverable, it's too short to predict everything."[19] Once again, Hawking had reversed his opinion on a fundamental problem in physics, and was not afraid to state that he had been mistaken.

IN THE PUBLIC EYE

Hawking was soon in the news once more, not because of a mistake he had made, but because of the misspoken statement of another physicist, Peter Higgs, the quiet, unassuming originator of the concept of the Higgs particle, responsible for making fundamental particles massive rather than massless in the standard model.[20] In early September 2002, Higgs was attending a play in Edinburgh based on the work of Paul Dirac and at an official dinner made some comments which took on a life of their own in the local press. Of Hawking, Higgs was reported to have said that it was "very difficult to engage him in discussion, so he has got away with pronouncements in a way that other people would not. His celebrity status gives him instant credibility that others do not have."[21] Other British newspapers picked up the story, sensationalizing it as much as possible. An anonymous physicist was quoted as saying "To criticize Hawking is a bit like criticizing Princess Diana—you just don't do it in public."[22]

Reporters were quick to point out that Hawking had made another of his celebrated bets against the discovery of the Higgs particle by the Large Electron Positron (LEP) experiment at CERN in Geneva. In 1996 Hawking had published a paper[23] in which he theorized that pairs of tiny virtual black holes should be produced, in analogy to the virtual particles constantly created out of the vacuum. These temporary black holes would "attract particles and scatter them back outwards." The net result, according to Hawking, was that the Higgs particle would become impossible to observe.[24] When the LEP experiment closed down in November 2000, without definitive proof of the long-sought Higgs particle, Hawking had collected $100 from colleague Gordon Kane of the University of Michigan. (A second bet stands unresolved that a similar experiment, the Tevatron, at Fermilab near Chicago, will also fail to find the Higgs particle.) Hawking was asked for his comments on the matter, and offered that he was "surprised by the depth of feeling in Higgs' remarks. I would hope one could discuss scientific issues without personal attacks."[25] In a later, unrelated interview, Higgs revealed that he had quickly contacted Hawking in private and explained the context of his comments. For his part, Hawking had responded that he was not offended, but had affirmed that he still felt that the Higgs particle might be beyond the reach of all experiments to directly detect.[26]

On a more humorous note, in the spring of 2003 Hawking was contacted by famed comedian Jim Carrey, asking if he would take part in a comedy sketch for the *Late Night with Conan O'Brien* television show. Hawking eagerly accepted, and taped webcam footage for the comedian. On the show, Carrey began discussing cosmology, and on cue Hawking called him on a cellphone, admonishing Carrey not to bother because "their pea brains cannot possibly grasp the concept." He soon after excused himself, saying that he had to go because he was watching Carrey's movie *Dumb and Dumber*, and "marveling at the pure genius of it."[27] Hawking later hosted the funny man at his home in Cambridge.

Hawking continued to keep a busy travel schedule that would tax the constitution of many able-bodied persons. He spent a month at the Mitchell Institute for Fundamental Physics at Texas A&M University, attending a scientific conference as well as delivering three public lectures. Directly afterward in late March he attended the Davis Meeting on Cosmic Inflation at the University of California, Davis, and gave another public lecture. In August he received the Oskar Klein Medal from the Royal Swedish Academy of Sciences, an annual honor given to a distinguished researcher. While in Sweden, he participated in the Nobel Symposium on String Theory and Cosmology and gave a public lecture at Stockholm University. Afterward, Hawking jetted back to America for another two-month stint at Caltech, the University of California–Santa Barbara, and Case Western Reserve University in Cleveland. At Case Western, he gave the Michelson-Morley Award lecture, one of the university's highest honors, recognizing "distinguished scientific achievement and contributions to the advancement of knowledge and improvement of human welfare." Could Hawking's health keep up with the demands on his time? The answer was, unfortunately, no. In December 2003, Hawking was admitted to Addenbrook's Hospital with pneumonia, and remained there for several weeks. He was readmitted in February 2004, after suffering a relapse, and had to cancel another planned trip to Texas in March.

While her father was out of the public eye for several months, Lucy was beginning to emerge into it. After working as a reporter for many years, Lucy had completed her first novel, *Jaded*, and was embarking on the media interview circuit in Britain. Along with questions about her book, there were the inevitable queries about her famous father. She replied that these questions were "another instance of feeling very overshadowed by him." Even though she was proud of her book, she said that "the interest is in what I can say about Stephen Hawking. That's not easy."[28] She explained that the book had been a diversion from the problems of her personal life.

After giving birth to her son, William, she had married, and then separated from, Alex Mackenzie Smith, William's father. Not long afterwards, when William was three years old, he had been diagnosed with autism. Upon receiving the news, Lucy said "I actually felt as if my heart was breaking into tiny pieces." Like her mother before her, she was faced with a dire prognosis for someone she dearly loved. It was her mother who had urged her to find a pediatrician who could offer William effective treatment, which has had a tremendous positive impact on the boy. Lucy credited her mother with teaching her a valuable lesson of life—to have "the determination to persevere, to find a way of doing something."[29] Stephen is proud of his grandson, and included a picture of the adorable toddler in *The Universe in a Nutshell.* William is likewise proud of his famous grandfather, bragging that his "grandad has wheels."[30] Lucy announced during an April radio interview that she was beginning to work on her second novel, tentatively entitled *My Glorious Breakdown.*

Hawking's return to the public eye was in dramatic form, literally. The BBC debuted its television drama *Hawking* in April 2004. The movie, directed by Philip Martin of *Stephen Hawking's Universe*, retold those two pivotal years in Hawking's life when he was diagnosed with ALS, met Jane Wilde, and conducted his thesis work on singularity theorems. Hawking had input into the final version of the script, which was partially based on Jane's autobiography. Actor Benedict Cummerbatch was uncanny as Stephen, the result of careful study of the early stages of ALS. Jane noted that his performance "brought back that period so very strongly." However, she admitted that the film did take some liberties with their lives. It was partially in fear of this inevitability that she had written her book in the first place, in case someone came back later to invent their lives. Nevertheless she believed that the film was true to the spirit of their early years together, and said it captured the "sense we had that, despite it all, everything was going to be possible."[31] The BBC 4 reran a thirty-minute documentary in concert with the movie, *Stephen Hawking: Profile*, which had been originally broadcast in 2002. *Hawking* proved popular, bringing in an estimated four million viewers.

With his health stabilized, it was announced early that summer that Hawking had teamed up with Silicon Graphics (a collaborator on the Cambridge COSMOS supercomputer) to create *Stephen Hawking's Beyond the Horizon*, a large-format IMAX film, scheduled for release in 2005, the 100th anniversary of Einstein's special theory of relativity and a worldwide celebration called The World Year of Physics. In Hawking's words, this popular-level film aimed to "bring the universe to everyone—from children to adults—around the world."[32]

PARADOX SOLVED?

Hawking's views of the universe still included the troubling opinion that information was irrevocably lost in black holes. Many of Hawking's fellow gravitational physicists agreed with his conclusion, while many particle physicists believed that a hint of a solution had already been spied in superstrings/M-theory. Given the fact that Hawking's "only vice is that he occasionally gambles,"[33] it came as a surprise to no one when Kip Thorne and Hawking made a bet against string theorist John Preskill on February 6, 1997, concerning a resolution of the black hole information problem. Preskill bet that the "mechanism for the information to be released by the evaporating black hole must and will be found in the correct theory of quantum gravity," the stakes being "an encyclopedia of the winner's choice, from which information can be recovered at will."[34]

Tantalizing rumors began circulating in the spring of 2004 that Hawking had derived his own solution to the information-loss paradox. He finally gave a seminar at Cambridge on his new theory, and some details began to emerge in the scientific press. Defying convention, he made a last minute request to Curt Cutler, the Chair of the Scientific Committee of the prestigious, week-long Seventeenth International Conference on General Relativity and Gravitation (GR17) in Dublin, asking to present his new theory without releasing an advanced copy of the article (or preprint). Cutler agreed, admitting to the press that he "went on Hawking's reputation."[35] Media coverage of Hawking's most recent admission of previous error was intensive and worldwide, and his talk on July 21 was not only packed with the six hundred physicists attending the meeting, but dozens of reporters as well. Conference organizer Petros Florides introduced Hawking, joking that even though it is believed that "no information can travel faster than light, this seems to have been contradicted by the speed with which the announcement of Hawking's talk spread around the globe."[36]

Hawking's talk invoked the anti-de Sitter/Conformal Field Theory (AdS/CFT) conjecture from brane theory,[37] as well as his usual mathematical trick of the Euclidean approach (using imaginary time) to the Feynman sum over histories. His results suggested that black holes could have more than one topology (geometry) at the same time, meaning that a true event horizon does not form, and hence information is not trapped. Since there is no need to explain where the information goes, there is no need for a baby universe branching off the black hole. Hawking apologized to science fiction fans, but firmly stated that "there is no possibility of using black holes to travel to other universes. If you jump into a black hole, your mass-energy will be returned to our universe, but in a mangled form, which contains the information about what you were like, but in an unrecognizable form."[38] After his talk, he invited Kip Thorne and John Preskill up onto the stage and formally conceded his part of the bet, presenting

Preskill with a baseball encyclopedia. Thorne chose not to concede just yet, wanting to wait until he could see more details in a promised upcoming article.[39] Preskill was pleased with his prize, but was concerned that he and Hawking would no longer have anything to argue about. He also admitted, "I didn't understand the talk," when asked about the new theory.[40]

Many colleagues agreed with Thorne and Preskill's assessment of the general overview Hawking had given of his new theory. Some were generally suspicious of the Euclidean technique, noting that it has yet to be proven that real time and imaginary time calculations always give rise to the same physical results. Also, Hawking did not discuss any details about how the information would be returned by the black hole, whether all at once or in a trickle.[41] Gerard 't Hooft was vocally "very disappointed with his 'final explanation,' "[42] while William Unruh of the University of British Columbia was more diplomatic:

> Part of the problem is he's providing so few details, so it's impossible to know whether we can believe these calculations. Stephen Hawking's not stupid, so we're going to take what he says seriously...but the whole theory we're hearing seems extremely speculative.[43]

Until a peer-reviewed paper on the theory is finally published, Hawking's colleagues, along with the rest of the world, will have to wait and see if he has once again proven himself to be the master of black holes.

NOTES

1. Stephen W. Hawking, *The Universe in a Nutshell* (New York: Bantam Books, 2001), p. vii.
2. Ibid.
3. Joseph Silk, "The Reality of a Brane New World," *Physics World* (December 2001): 43.
4. Hawking, *The Universe in a Nutshell*, p. 195.
5. Ibid., p. 54.
6. *PhysicsWeb*, "Hawking Scoops Book Prize," June 26, 2002, http://physicsweb.org/article/news/6/6/14/1 (accessed August 27, 2004).
7. *BBC News*, "Hawking Takes Top Book Prize," June 25, 2002, http://news.bbc.co.uk/1/hi/sci/tech/2066033.stm (accessed September 8, 2003).
8. Roger Highfield, "Hawking's Triumph Over Time," *The Telegraph*, January 8,

2002, http://www.telegraph.co.uk/news/main.jhtml?xml=/news/2002/01/08/nhawk08.xml (accessed March 18, 2003).

9. Peter Rodgers, "From the Big Bang to the Eureka Moment," *Physics World* (February 2002): 9.

10. Roger Penrose, "The Problem of Spacetime Singularities: Implications for Quantum Gravity?" in *The Future of Theoretical Physics and Cosmology,* ed. G. W Gibbons, E. P. S. Shellard, and S. J. Rankin (Cambridge: Cambridge University Press, 2003), p. 51.

11. Lisa Sewards, "A Brief History of Our Time Together," *Daily Telegraph,* April 27, 2002, http://www.lexisnexis.com (accessed August 19, 2004).

12. Tim Radford, "A Life, the Universe, and a Big Bang," *The Guardian,* January 12, 2002, http://www.guardian.co.uk/uk_news/story/0,,631614,00.html (accessed August 27, 2004).

13. *BBC News,* "Hawking Extols Joy of Discovery," January 11, 2002, http://news.bbc.co.uk/1/hi/sci/tech/1755683.stm (accessed August 19, 2004).

14. *Society of Academic Authors,* "Author Objects to Repackaging of His Work," May 24, 2002, http://www.sa2.info/ARCHIVE/2002/05mayB.html (accessed August 27, 2004).

15. Stephen Hawking, "Information," *Professor Stephen Hawking's Web Pages,* http://www.hawking.org.uk/info/iindex.html (accessed August 19, 2004).

16. Stephen Hawking, ed., *On the Shoulders of Giants* (Philadelphia: Running Press, 2002).

17. Gregory Benford, "Leaping the Abyss," *Reason Online,* April 2002, http://reason.com/0204/fe.gb.leaping.shtml (accessed September 8, 2003).

18. Stephen Hawking, "Gödel and the End of Physics," *Dirac Centennial Celebration,* July 20, 2002, http://www.damtp.cam.ac.uk/strtst/dirac/hawking (accessed August 27, 2004).

19. James Hartle, "Theories of Everything and Hawking's Wave Function," in *The Future of Theoretical Physics and Cosmology,* ed. G. W. Gibbons, E. P. S. Shellard, and S. J. Rankin (Cambridge: Cambridge University Press, 2003), p. 49.

20. The Higgs field is discussed in Appendix C on the details of inflation theory.

21. Alastair Dalton, "Clash of the Atom-Smashing Academics," *The Scotsman* (September 2, 2002): 3.

22. Steve Connors, "Higgs vs. Hawking: A Battle of the Heavyweights That Has Shaken the World of Theoretical Physics," *The Independent,* September 3, 2002, http://millennium-debate.org/ind3sept023.htm (accessed August 19, 2004).

23. S. W. Hawking, "Virtual Black Holes," *Physical Review D* 53 (1996): 3099.

24. John Gribbin, "Hawking Throws Higgs Into Black Holes," *New Scientist* (December 2, 1995): 20.

25. Donald MacLeod, "Physicists Embroiled in Celebrity Row," *Guardian Unlim-*

ited, September 3, 2002, http://education.guardian.co.uk/higher/news/story/0,9830,785503,00.html (accessed August 27, 2004).

26. Peter Rodgers, "Peter Higgs: The Man Behind the Boson," *Physics World* (July 2004): 11.

27. Jim Carrey, interview by Andrew Denton, *Enough Rope with Andrew Denton*, Australian Broadcast Corporation, June 16, 2003.

28. Elizabeth Grice, "Dad's Important, But We Matter, Too," *The Telegraph*, April 13, 2004, http://www.telegraph.co.uk/arts/main.jhtml?xml=/arts/2004/04/13/bohawk13.xml (accessed May 25, 2004).

29. Emine Saner, "A Brief History of Mine," *The Scotsman*, April 20, 2004, http://lifestyle.scotsman.com/living/headlines_specific.cfm?articleid=8344 (accessed May 25, 2004).

30. Grice, "Dad's Important, But We Matter, Too."

31. Tim Adams, "Brief History of a First Wife," *The Observer*, April 4, 2004, http://observer.guardian.co.uk/review/story/0,,1185067,00.html (accessed August 27, 2004).

32. *E.E. Product Center*, "Silicon Graphics and Universe Partners, LLC, Collaborate on Upcoming Film 'Stephen Hawking's Beyond the Horizon,'" June 3, 2004, http://www.eeproductcenter.com/showpressrelease.jhtml?articleID=x214376 (accessed August 27, 2004).

33. Malcolm Perry, "Black Holes and String Theory," in *The Future of Theoretical Physics and Cosmology*, ed. G. W. Gibbons, E. P. S. Shellard, and S. J. Rankin (Cambridge: Cambridge University Press, 2003), p. 295.

34. Ibid., p. 296.

35. Jenny Hogan, "Hawking Cracks Black Hole Paradox," *New Scientist* (July 17, 2004): 11.

36. John Baez, "Week 207," *This Week's Finds in Mathematical Physics*, July 25, 2004, http://math.ucr.edu/home/baez/week207.html (accessed September 5, 2004).

37. The AdS/CFT conjecture is discussed in Appendix D.

38. Baez, "Week 207."

39. *New Scientist*, "Holey Let-down" (July 31, 2004): 4.

40. Associated Press, "Physicist Rethinks Theory on Black Holes," *ABC News*, July 22, 2004, http://abcnews.go.com/wire/World/ap20040721_1674.html (accessed July 28, 2004).

41. Charles Seife, "Hawking Slays His Own Paradox, But Colleagues Are Wary," *Science* 305(2004): 586.

42. Peter Rodgers and Matin Durrani, "Hawking Loses Wager on Black Holes," *Physics World* (August 4, 2004): 5.

43. Associated Press, "Physicist Rethinks Theory on Black Holes."

EPILOGUE

STEPHEN HAWKING

Man vs. Myth

What are we finally to make of Dr. Stephen Hawking, world-famous physicist and cultural icon? Over the past four decades, he has sparked serious scientific debate and provided the impetus for major discoveries, both his own and those of colleagues. He has served as an ambassador for science, bringing the universe a little more down to earth for the common person. He has also been a positive and vocal role model for those with physical challenges. But in the end, he is most fundamentally a colleague, friend, mentor, father, grandfather, and husband.

If he never publishes a single scientific paper again, his legacy will be secured for decades, or perhaps even centuries, to come. But even with his numerous awards and twelve honorary degrees, one honor has continued to elude him—a Nobel Prize. Such a distinctive honor awaits some experimental verification of the ground breaking theories that the bottomless intellectual spring of his mind continually seems to bring forth. Hawking himself keeps it in perspective: "It is better to go on and make new discoveries than to hope for a prize for work I did years ago."[1] Close friend and long-time collaborator Jim Hartle succinctly stated that Hawking's "remarkable combination of boldness, vision, insight, and courage have enabled him to produce ideas that have transformed our understanding of space and time, black holes, and the origin of the universe."[2]

Hawking's distinctive personality certainly cannot be ignored. Just as Leonard Susskind called him infuriatingly stubborn, the late David Schramm painted his friend as "an incorrigible flirt: a party animal who likes to dance in his wheelchair." People who annoy him risk having their toes run over, and he once rammed a car with his wheelchair because the vehicle was illegally blocking his ramp. In typical Hawking humor, he offered that such stories are "a malicious rumor. I'll run over anyone who repeats it."[3] He joked that he once had thoughts of a political career, perhaps even the job of Prime Minister. Hawking said, "I was glad I left that job to Tony Blair. I think I get more job satisfaction than he does and I expect my work will last longer."[4] When asked about his views on science fiction, he remarked that "I write it, only I like to think it is science fact."[5]

Visitors to his office are greeted with such irreverent decorations as a Homer Simpson clock and a photo-manipulation of Hawking with Marilyn Monroe. His proclivity for making research-related bets has recently taken on a life of its own. Bookmakers Ladbrokes opened up bets for a two-week period in August 2004 on five ongoing and future physics projects, including experimental verification of the Higgs particle. It was proudly announced that "[b]etting on the greatest unsolved problems in the universe is no longer the preserve of academic superstars such as Stephen Hawking."[6]

The unmistakable image of his computer-equipped electric wheelchair and the sound of his mechanical voice are forever burned into our collective consciousness. When a tourist to Cambridge stopped him in the street and asked if he was the famous Stephen Hawking, he replied, "the real one was much better looking!"[7] His undeniable status as a pop icon has not always been so good-natured, however. The unrelenting intrusions of the press into his personal life, especially since his separation from Jane, have been difficult, and as a result he speaks even less about his family life than he did during his first marriage. He does realize that although the lack of privacy is annoying, it is an unavoidable part of fame, and said "it would be hypocritical to complain. I can generally ignore it by going off to think in eleven dimensions."[8]

Hawking has repeatedly said that pronouncements of his great courage as a disabled person are embarrassing. He countered that "I have only done what I intended to do anyway, before I had ALS. I think the people with real courage are those worse affected but who don't get public attention or sympathy. Yet they don't complain."[9] Despite his self-deprecating remarks, he remains a very visible symbol of success and hope for those facing serious physical challenges. He said he willingly accepts that "there are some things I can't do," but notes, "they are mostly things I don't particularly want to do anyway."[10] He is openly thankful that his disease has allowed him "to concentrate on research without having to lecture or sit on boring committees."[11] *Star Trek* fan Diane Smith, whose own mother died of ALS after a six-year struggle with the disease, eloquently summarized the universality of Hawking's appeal:

> Who would presume that I, with only two years of college, could simply buy a ticket and sit in the august presence of so many eggheads, including possibly the smartest man on earth? How many PhDs are here, anyway? However, a large part of the audience seems to be from all walks of life—children, elders, Aggies, Trekkers—and I feel I fit right in. All races, all creeds are here—just like a *Trek* convention. Everyone is quiet, dignified, and hanging on his every word. We laugh when he tells a joke and we applaud when his truths strike us.[12]

Hawking continues to remind us why such esoteric studies as singularity theorems and M-theory are worthy realms of human inquiry. He says they may not "help feed anyone or get their wash whiter. But men and women do not live by bread alone. We all need to understand where we come from and these observations show us a glimpse of our origin."[13] Hawking embodies the very spirit of scientific exploration and the self-correcting nature of science, where admitting error is a normal sign of progress rather than shame. Perhaps the true nature of Stephen Hawking is best embodied in his happiness over the fact that "our search for understanding will never come to an end, and that we will always have the challenge of new discovery. Without it, we would stagnate."[14]

That is Stephen Hawking "in a nutshell."

NOTES

1. Michael Lemonick, "Hawking Gets Personal," *Time* (September 27, 1993): 80.

2. Gary Gibbons and Paul Shellard, "Introduction," in *The Future of Theoretical Physics and Cosmology*, ed. G. W. Gibbons, E. P. S. Shellard, and S. J. Rankin (Cambridge: Cambridge University Press, 2003), p. 1.

3. Nigel Farndale, "A Brief History of the Future," *Sydney Morning Herald*, January 7, 2000, http://www.smh.au/news/0001/07/features/feature1.html (accessed March 27, 2003).

4. Stephen Hawking, interview by Larry King, *Larry King Live Weekend,* December 25, 1999.

5. *The New Yorker* , "A Brief History" (April 18, 1988): 31.

6. Valerie Jamieson, "Biggest Bets in the Universe Unveiled," *New Scientist,* August 7, 2004, http://www.newscientist.com/news/news.jsp?id=ns99996331 (accessed September 23, 2004).

7. Bernard Carr, "Primordial Black Holes, in *The Future of Theoretical Physics and Cosmology,* ed. G. W. Gibbons, E. P. S. Shellard, and S. J. Rankin (Cambridge: Cambridge University Press, 2003), p. 259.

8. Farndale, "A Brief History of the Future."

9. Russ Sampson, "Two Hours with Stephen Hawking," *Astronomy* (March 2003): 16.

10. Ron Kampeas, "Stephen Hawking, Wired for the Web, is More Switched on Than Ever," *Augusta Chronicle Online,* March 23, 1997, http://augustachronicle .com/stories/032397/tech_hawking.html (accessed November 10, 2001).

11. Stephen Hawking, *Larry King Live Weekend.*

12. Diane Smith, "An Evening with Stephen Hawking," *Trek Nation,* March 24, 2003, http://www.treknation.com/articles/an_evening_with_stephen_hawking. shtml (accessed August 27, 2004).

13. Sampson, "Two Hours with Stephen Hawking": 16.

14. Stephen Hawking, "Gödel and the End of Physics," *Dirac Centennial Celebration,* July 20, 2002 http://www.damtp.cam.ac.uk/strtst/dirac/hawking (accessed August 27, 2004).

AFTERWORD

Stephen Hawking has continued to fascinate and educate us in the two and a half years since I completed the manuscript for the first edition of this book. Not surprisingly, several more distinguished awards have come his way. On Valentine's Day 2005, Hawking was feted by the Smithsonian Institution in Washington, DC. A retrospective of his life was presented by friend and colleague James Hartle, and Hawking himself offered a few remarks. The main point of the occasion was to award Hawking the James Smithson Bicentennial Medal, given in honor of distinguished contributions to areas of interest to the institution. Ten days later, Hawking honored his former mentor by delivering the third Dennis Sciama Memorial Lecture at Oxford. The most prestigious honor came on November 30, 2006, when he was awarded the Copley Medal by the Royal Society, the august society's highest award, presented for outstanding research achievements. This award was particularly personal, as the silver medal had been carried into space five months before by British astronaut Piers Sellers on a space shuttle mission to the International Space Station. Sellers explained that

> Stephen Hawking is a definitive hero to all of us involved in exploring the cosmos. His contribution to science is unique and he serves as a continuous inspiration to every thinking person. It was an honor for the crew of the STS-121 mission to fly his medal into space. We think that this is particularly

appropriate as Stephen has dedicated his life to thinking about the larger universe.[1]

But as we have certainly seen, scientists are not the only ones to appreciate the impact Hawking has had over the past few decades. For example, he took second place in a 2004 poll of top role models which surveyed five hundred sixteen- to eighteen-year-old English boys. Although he beat out third-place winner soccer star David Beckham, he was bested by another athlete, rugby player Jonny Wilkinson.[2] Hawking nearly stole the show at the 2004 British Comedy Awards when he presented an award to Matt Groening, creator of *The Simpsons.* Show host Jonathan Ross admitted that he had never been able to read all of Hawking's famed best seller *A Brief History of Time,* and asked if it had a happy ending. Hawking's cheeky response—since it had sold over two hundred million copies, it certainly had a happy ending for him![3]

Hawking reprised his role on *The Simpsons* in May 2005, in the episode "Don't Fear the Roofer." At a surprise birthday party for Lenny in Moe's bar, Hawking's autobiographical character announced that he was now a Springfield resident, having purchased the neighborhood's Little Caesar's pizza franchise. When his attempt to say the company motto resulted in a runaway refrain of "pizza pizza," the cartoon version of Hawking smacked his computer and announced with chagrin, "sorry, that button sticks." Later in the episode, Hawking helped preserve Homer's sanity by explaining that no one but Homer could see Ray Romano's character in Builder's Barn due to a tear in space-time that created a small black hole, resulting in gravitational lensing.

Hawking's long-standing status as a pop icon was further acknowledged in the past year. In December 2006, it was announced that the producers of the British reality show *Celebrity Big House* had offered Hawking a contestant slot on the show. In February 2007, two references to Hawking and his work appeared in separate episodes of the popular and intellectually intricate ABC television show *Lost.* In the episode "Not in Portland," the minor character Aldo was seen reading *A Brief History of Time.* Detail-oriented fans noted the exact pages from the chapter "Black Holes Ain't So Black" being shown, and pondered in online chatrooms and discussion boards the significance of the reference to the show's unfolding plotline. The surprising answer came in the next episode, "Flashes before Your Eyes," where recurring character Desmond Hume appeared to travel back in time through some sort of space-time anomaly created by a large implosion he had generated. When he attempted to change the past, he was rebuked by a mysterious jewelry store owner named Ms. Hawking.

Other television appearances have recently come Hawking's way. He appeared as an expert consultant in the Discovery Channel docudrama *Alien Planet* in 2005. The *Hollywood Reporter* announced in August 2006 that Hawking was tapped to introduce each episode of a science fiction anthology show under production titled *Masters of Science Fiction*. *Hawking*, the acclaimed BBC drama about the scientist's early career, finally debuted on American television in January 2007, on the Science Channel.

Hawking has also continued in his efforts to bring science to a general reading audience. The illustrated version of his *On the Shoulders of Giants* appeared in October 2004, and a companion volume, *God Created the Integers*, was published a year later. Similar to *On the Shoulders of Giants*, the second book was a compilation of historically important works, but of mathematics (largely mathematical proofs) rather than physics. As with the first volume, this work included Hawking's summaries of the mathematicians' lives. In response to long-standing criticisms concerning the technical level of certain parts of his most famous popular-level book, Hawking and coauthor physicist Leonard Mlodinow published *A Briefer History of Time* in October 2005. Slightly shorter than the original best seller and illustrated with appealing color images, this new book was updated to reflect current scientific understanding, and more importantly avoided detailed discussion of some of the more counterintuitive topics, including the no-boundary condition. Reviews were mixed, but Hawking's intentions were accurately summed up in one particular review by physicist and author of popular-level science books Jim Al-Khalili in the science journal *Nature*: "I believe the authors have made an honest attempt here to rectify what they perceive as a problem with the original: that millions of readers with no scientific background did not get beyond the first chapter before their brains blew up."[4] Mlodinow and Hawking are currently working on two other projects, the IMAX movie *Beyond the Horizon* and *The Grand Design*, a book on the origin of the universe and the basis of the laws of physics, due to be published in late 2008.

As we have previously seen, writing is a family affair for the Hawkings, as daughter Lucy is a journalist and novelist. During a six-day trip to Hong Kong in June 2006, Stephen and Lucy announced that they would be publishing a children's book on science which would be "a bit like Harry Potter, but without the magic."[5] *George's Secret Key to the Universe*, featuring the adventures of

George and his neighbors, the scientist Eric, daughter Annie, and a supercomputer named Cosmos, is slated to appear in September 2007, with two other books possibly to follow. Lucy's second novel was also published soon after the announcement, titled *The Accidental Marathon* in the UK and *Run for Your Life* in the United States.

Any realistic update on the life and work of the complex human being who is Stephen Hawking must include his sometimes political and controversial public statements. In a June 2006 seminar in Beijing, China, Hawking amused his audience by lauding the beauty of Chinese women. But his comments turned more serious when asked about the environment. He affirmed his serious concern about global warming, and worried that our planet "might end up like Venus, at 250 centigrade and raining sulfuric acid."[6] In the summer of 2006, Hawking posed the following question to online readers of Yahoo! Answers: "In a world that is in chaos politically, socially and environmentally, how can the human race sustain another 100 years?" Twenty-five thousand responses were received. In his own posted follow-up, he admitted that he didn't have an answer, which is "why I asked the question, to get people to think about it, and to be aware of the dangers we now face." He did offer that perhaps "we must hope that genetic engineering will make us wise and less aggressive."[7] Hawking reaffirmed his long-standing fears for the future of the human race in a November 2006 BBC radio interview, where he made the widely reported comment that "the long-term survival of the human race is at risk as long as it is confined to a single planet.... There isn't anywhere like the Earth in the solar system. For that, we have to go to another star."[8]

Hawking's concerns about both the environmental and nuclear Armageddon were combined in his January 2007 appearance at the Royal Society at the occasion of the *Bulletin of Atomic Scientists'* Doomsday clock being reset. Taking into account the current state of the world, the clock was set two minutes closer to midnight, which is defined on the clock as Doomsday. Commenting on the clock's resetting to 11:55 PM, Hawking warned:

> As scientists we understand the dangers of nuclear weapons and their devastating effects, and we are learning how human activities and technologies are affecting climate systems in ways that may forever change life on Earth.... As citizens of the world, we have a duty to share that knowledge, and to alert the public to the unnecessary risks that we live with every day. We foresee great peril if governments and societies do not take action now to render nuclear weapons obsolete and to prevent further climate change.[9]

Stephen also lent his considerable presence to a January 2007 petition signed by a hundred scientists, religious leaders, actors, writers, and members of British Parliament in protesting the planned replacement of the Trident nuclear weapon system. In a statement to the London paper the *Independent*, Hawking warned that "[n]uclear war remains the greatest danger to the survival of the human race. To replace Trident would make it more difficult to get arms reduction and increase the risk."[10]

The Iraq War also drew sharp criticism from the noted scientist. At a November 2004 antiwar rally in London's Trafalgar Square, Hawking condemned the American invasion of Iraq as a "warcrime," and afterwards added that it was "based on two lies. The first was we were in danger of weapons of mass destruction and the second was that Iraq was somehow to blame for Sept. 11."[11] His longtime disdain for politicians was also evident at a November 2005 public talk telecast to Seattle. When asked about President George W. Bush's plan to return astronauts to the moon, Hawking called the plan "stupid" and added that "sending politicians would be much cheaper, because you don't have to bring them back."[12] He also dismissed a July 2006 proposal by members of the European Union to ban stem cell research, countering that "[t]he fact that the cells may come from embryos is not an objection because the embryos are going to die anyway. It is morally equivalent to taking a heart transplant from a victim of a car accident."[13]

As controversial as some of these comments may be, it was a humorous quip integrated into an often repeated public talk on the origin of the universe that drew sharp criticism from one particular group. Referring to the talk he had given about preliminary work on the no-boundary proposal at the 1981 Study Week on Cosmology and Fundamental Physics at the Vatican (see chapter 7), Hawking recalled that he was glad that Pope John Paul II, who, he recalled, had admonished that scientists should not study the origin of the universe, was unaware of the topic of his talk, because he "didn't fancy the thought of being handed over to the Inquisition like Galileo."[14] Although the comment had been repeated at numerous public talks, and had appeared in the press at least once before, after wide press coverage of a June 2006 sold-out talk in Hong Kong, the remark elicited a rather heated response from Catholic League president Bill Donohue:

> There is a monumental difference between saying that there are certain questions that science cannot answer—which is what the pope said—and

> authoritarian pronouncements warning scientists to back off.... Hawking, who claims—without any evidence—that space and time have no beginning and no end, would be wise to refrain from positing false absolutes and learn to realize when he's out of his league. Most important, he should stop distorting the words of the pope.

An article on the incident appeared in the Catholic League's journal, *Catalyst*, which complained that "Hawking got away with his little stunt because he's the darling of the media. They treat him as if he's some sort of saintly scientist who can do no wrong. Indeed, the same media outlets that ran with Hawking's erroneous account of what the pope said at the conference failed to do a follow-up story after we exposed his botched rendering of the facts."[15]

Hawking may or may not be a darling in the eyes of the media, but his words often do bring swift response. Case in point was a November 2006 BBC radio interview in which he stated that "my next goal is to go into space. Maybe Richard Branson will help me."[16] Within a few days, the following official statement was e-mailed to media outlets from the office of British entrepreneur Sir Richard Branson of Virgin Airlines fame:

> Obviously we would be honoured to have Stephen fly with Virgin Galactic. We have a great medical team and we are planning to have our Chief Medical Officer sit down with Stephen, and we will do everything in our power to make his dream of going to space possible.[17]

Virgin Galactic is the private spaceflight company founded by Branson, which plans to offer suborbital spaceflights beginning in late 2008 or 2009 for approximately $200,000 a seat. The fee was waived in Hawking's case. In preparation for his planned 2009 spaceflight, Hawking has accepted a free weightless experience in a 747 airplane offered by the company Zero Gravity. Company founder Peter Diamond is the chairman of the X Prize Foundation, which has developed a $10 million Archon X genomics prize for the first scientific team that can "build a device and use it to sequence 100 human genomes within 10 days or less with an accuracy of no more than 1 error in 100,000 base pairs... at a demonstrated cost of no more than $10,000 per genome."[18] Hawking has not only reportedly offered his DNA for analysis as part of the competition, but is quoted on the competition Webpage as a supporter of the project, saying in part, "It is my sincere hope that the Archon X PRIZE for Genomics can help drive breakthroughs in diseases like ALS at the same time that future X PRIZEs for space travel help humanity to become a galactic species."[19]

As previously mentioned, during the last several years Hawking has continued to attend professional conferences and give public talks around the world. For example, in 2005 he spent January at Caltech and University of California–Santa Barbara, where he introduced a talk by his friend and Nobel Prize laureate David Gross. March found him in Spain celebrating the twenty-fifth anniversary of the Prince of Asturias awards, June brought six days of talks in Hong Kong as part of a longer trip to a String Conference in Beijing, and October found him in Germany to speak at a university in Berlin and attend a book fair in Frankfurt (to coincide with the publication of *A Briefer History of Time*).[20] November 2005 brought him back to the United States, with three public talks on the origin of the universe (which included the controversial line discussed above) scheduled for San Jose and Oakland, California, and Seattle, Washington. But health problems interfered with the end of his 2005 travel schedule. Before traveling from Oakland to Seattle, there was an incident while he was being taken off his respirator in the morning. According to the host of the Seattle appearance, Hawking "basically flat-lined. They had to resuscitate him, and that panicked a few people. But he's been there before."[21] It was decided to err on the side of caution, and Hawking delayed further travel, delivering the Seattle talk by live telecast instead of a live personal appearance. Other travels not mentioned previously include a December 2006 tour of Israel and the Palestinian territories, where he met with students, scientists, politicians, and of course, the media. In an interview with Israeli talk-show host Yair Lapid, he offered that the upside of his condition was that he was largely exempted from "boring committees," while the downside to his celebrity was his lack of anonymity when traveling. As he explained, "It is not enough for me to wear dark sunglasses and a wig. The wheelchair gives me away."[22]

For the past four decades, ALS has been Hawking's constant companion, and although his longevity and ability to thrive both personally and professionally despite the disease are impressive, the disease continues to make it difficult for the scientist to communicate with others. Since 2000, Stephen had found that the hand he used to control his computer was becoming weaker and his ability to use his traditional mode of communication was becoming frustratingly slower.[23] In response, a new communication technology began to be utilized. It was announced in 2005 that Hawking was now using the

Infrared/Sound/Touch (IST) switch developed by Words+. The device is attached to his eyeglasses, and its low-power infrared beam is controlled by blinking an eye or otherwise moving a cheek muscle.[24]

And what of Hawking's personal life, which had been the source of a decade and half of continuous speculation since his separation and divorce from his first wife, Jane Wilde, and his marriage to Elaine Mason, his former nurse? Despite his repeated attempts to maintain a normal and private home life, the media continued to hound him, sniffing out any possible hint of scandal. In the summer of 2006, the author of this volume received a phone call from a London tabloid, asking if I had any knowledge about problems in the Hawkings' marriage. The call dismayed me on several levels. Of course, the thought that Hawking's second marriage might be in serious trouble was disconcerting, but certainly not shocking, given the nearly continuous rumors which had been circulating throughout the previous decade. What was harder for me to fathom was that the media was so determined to dig up dirt on what would be, if true, an extremely unhappy episode of Professor Hawking's life, that they would feel it was worth contacting an American author who had published a biography of the scientist. It was therefore no surprise to me to read in October 2006 that Stephen and Elaine's marriage had succumbed to some unknown combination of internal and/or external pressures and that they had filed for divorce.[25]

By now you must be wondering why I haven't discussed Hawking's scientific publications during the past few years. Besides an admittedly perverse desire to make the reader wait as long as possible to learn more about Hawking's controversial resolution of the black hole information paradox, it is important to set his work in the larger context of the pressures, both personal and physical, which have surrounded Hawking since presenting the initial results in July 2004. As noted in chapter 12, Hawking suffered a serious bout of pneumonia in late 2003 through early 2004. What we now know, courtesy of the BBC Horizon documentary *The Hawking Paradox* (which first aired in September 2005), is that for eighteen months prior to the illness, Hawking and student Christophe Galfard had been looking at the problem through the lens of the AdS/CFT correspondence (see appendix D) with little success. It was during his illness that Hawking had the breakthrough of insight which resulted in the solution he presented at GR17 in July 2004. As discussed at the end of chapter 12, attendees at the conference were skeptical at best, confused at worst, by Hawking's overview of his solution. It was believed that Hawking would soon release a preprint version of a paper based on this solution at the same time it would be submitted for peer-reviewed publication. But as the months passed, no such preprint appeared in the professional preprint online archives. *The Hawking Paradox* ended with Galfard and Hawking working on the promised paper, explaining that "progress is tortuously slow," due to Hawking's increasing problems using his manual computer control.

Finally, on August 22, 2005, the paper arrived at the offices of *Physical Review D*, and after undergoing the peer-review process, was published in the October 18, 2005, edition of the journal. The crux of Hawking's argument is as follows. Using Feynman's sum over histories approach, Hawking took into account the contributions of possible configurations of the universe both with and without black holes. As Hawking wrote, "Information is lost in topologically nontrivial metrics like black holes... [but] information about the exact state is preserved in topologically trivial metrics," or in English, in the configurations with black holes, information will be lost, but in those without black holes, information will not be lost.[26] When the probabilistic contributions of all the possible configurations are summed, the contribution of those configurations without black holes is more significant, so overall information will be conserved. At three and a half pages and containing only three equations, the paper was still lacking sufficient details to satisfy some of Hawking's colleagues. For example, a preprint posted by Martin Einhorn of University of California–Santa Barbara the day after Hawking's paper was published complained that Hawking's solution "seems to throw the baby out with the bathwater, so his widely publicized 'concession' remains quite controversial."[27] Similarly a 2006 paper summed up the current status of the information paradox as follows: "While some may have changed their minds, few would argue that the situation is resolved, and the mystery is, if anything, more pronounced."[28]

While his colleagues continued to digest his latest contribution to the ongoing discussion of the black hole information paradox, Hawking and Thomas Hertog renewed their earlier collaboration (which had previously been on brane models), resulting in a 2006 paper[29] in *Physical Review D* that caught the attention of the scientific media. In their radical new approach, they combined the no-boundary proposal with the so-called landscape of string theory. One of the dirty little secrets that makes many string theorists uncomfortable with the current status of their discipline is that there are predicted to be as many as 10^{500} or more possible states for the universe. The accumulation of all such states is termed the landscape of string theory. Some physicists, such as Leonard Susskind, have applied the anthropic principle to the problem in an attempt to get a grip on the number of possible universes—clearly those states that do not match our observable universe in any significant way are not important to our consideration. In Hawking and Hertog's so-called top-down approach, the early universe is pictured as a superposition of all possible states from the landscape, with each state having its own subsequent history. Therefore the universe has a plethora of different histories, each having its own probability of occurrence. By observing the universe in its current state, the proposal is that one can then work backwards and determine those possible initial states which could lead to what is observed today. If this methodology is correct, it would lead to slight differences in both the gravity wave and the cosmic microwave background spectra as compared to the predictions of standard inflationary cosmology. Such differences are theoretically within the reach of future technology to measure.[30]

We have now reached the end of our update of the life of one extraordinary member of the human race. In an interview published on his sixty-fifth birthday (January 8, 2007), a landmark surely no one familiar with ALS could have foreseen he would reach, Hawking explained that although the retirement age at Cambridge is sixty-seven, "I shall continue working."[31] And why shouldn't he, for as Hawking has proved time and time again, the only true limit to the human mind is the depth of its imagination. We would all benefit from adhering to Hawking's philosophy of life, revealed in a November 2006 BBC radio interview:

> I am not afraid of death but I'm in no hurry to die. I have so much more I want to do.[32]

I look forward to writing an update on the next chapter of Stephen Hawking's life, sometime after he attains his goal of reaching space.

—Kristine Larsen, February 24, 2007

NOTES

1. NASA press release, "Stephen Hawking to Receive Medal Flown on Space Shuttle," November 28, 2006, http://www.spaceref.com/news/viewpr.html?pid=21367 (accessed February 21, 2007).

2. "Hawking Tops Becks in Boys' Poll," BBC News, October 6, 2004, http://news.bbc.co.uk/1/hi/england/3719128.stm (accessed January 5, 2007).

3. British Comedy Awards, "Top Ten Moments of British Comedy Awards," http://www.britishcomedyawards.com/topmoments.html (accessed December 29, 2006).

4. Jim Al-Khalili, "Shortcut to Space-Time," Nature 438 (2005): 159.

5. "Hawking to Write Children's Book," *BBC* News, June 13, 2006, http://news.bbc.co.uk/2/hi/entertainment/5075516.stm (accessed January 5, 2007).

6. Alexa Olesen, "Stephen Hawking: Earth Could Become Like Venus," June 22, 2006, http://www.livescience.com/environment/ap_060622_hawking_climate.html (accessed January 2, 2007).

7. Tara Kirchner, "And the Answer Is..." Yahoo! Searchblog, August 1, 2006, http://www.ysearchblog.com/archives/000336.html (accessed December 29, 2006).

8. Harry MacAdam, "Search Is Vital, Says Hawking," *Sun*, December 28, 2006, http://www.thesun.co.uk/article/0,,2-2006600196,00.html (accessed December 29, 2006).

9. Geoffrey Lean, "Prophet of Doomsday: Stephen Hawking, Eco-Warrior," *Independent*, January 21, 2007, http://www.countercurrents.org/cc-lean230107.htm (accessed January 22, 2007).

10. Colin Brown, "Not in Our Name: Campaign Launched against Trident," *Independent*, February 15, 2007, http://news.independent.co.uk/uk/politics/article2271662.ece (accessed February 23, 2007).

11. Associated Press, "Scientist Stepher Hawking Decries Iraq War," *USA Today*, November 3, 2004, http://www.usatoday.com/news/world/2004-11-03-hawking-iraq_x.htm (accessed January 2, 2007).

12. Alan Boyle, "The Show Goes On for Stephen Hawking," MSNBC, November 15, 2005, http://www.msnbc.msn.com/id/10086479 (accessed December 29, 2006).

13. Steve Connor and Stephen Castle, "Hawking Criticizes EU States Trying to Ban Stem Cell Research," *Independent*, July 24, 2006, http://news.independent.co.uk/world/science_technology/article1193119.ece (accessed January 2, 2007).

14. Associated Press, "Stephen Hawking Touches on God and Science," MSNBC, June 15, 2006, http://www.msnbc.msn.com/id/13340672 (accessed January 29, 2007).

15. "Hawking Misrepresents Pope John Paul II," *Catalyst* 31, no. 6 (2006) http://www.catholicleague.org/catalyst/2006_catalyst/07806.htm#broward (accessed January 29, 2007).

16. MacAdam, "Search Is Vital."

17. Alan Boyle, "Stephen Hawking in Space," *Cosmic Log*, November 30, 2006, http://cosmiclog.msnbc.msn.com/archive/2006/11/30/16569.aspx (accessed December 29, 2006).

18. Archon X Prize, "Competition Guidelines," November 2, 2006, http://genomics.xprize.org/assets/downloads/Archon_X_PRIZE_for_Genomics_Competiton_Guidelines.pdf (accessed February 23, 2007).

19. Archon X Prize, "Archon X Prize in Genomics," October 4, 2006, http://genomics.xprize.org/ (accessed February 23, 2007).

20. Stephen Hawking, "News," nd, http://www.hawking.org.uk.info/news.html (accessed December 29, 2006).

21. Boyle, "The Show Goes On for Stephen Hawking."

22. "Hawking's Humor," *Israel Today*, January 28, 2007, http://www.israeltoday.co.il/default.aspx?tabid=128&view=item&idx=1248 (accessed February 23, 2007).

23. "Hawking Adopts Blink Technology," *Cambridge Evening News*, September 5, 2005, http://www.cambridge-news.co.uk/news/city/2005/09/05/4197bac3-3dfb-4d25-8f49-a5629c6a85f9.lpf (accessed February 23, 2007).

24. "The IST Switch, Now Used by Stephen Hawking," *MedGadget*, September 6, 2005, http://www.medgadget.com/archives/2005/09/the_ist_switch.html (accessed December 19, 2006).

25. Martin Hodgson, "Hawking to Divorce Wife after 11 Years' Marriage," *Independent*, October 20, 2006, http://news.independent.co.uk/uk/this_britain/article1905016.ece (accessed January 2, 2007).

26. S. W. Hawking, "Information Loss in Black Holes," *Physical Review D* 72 (2005): 084013-3.

27. Martin B. Einhorn, "The Black Hole Information Paradox," arxiv:hep-th/0510148, October 19, 2005 (accessed January 5, 2007).

28. John A. Smolin and Jonathan Oppenheim, "Locking Information in Black Holes," *Physical Review Letters* 96 (2006): 081302-1.

29. S. W. Hawking and Thomas Hertog, "Populating the Landscape: A Top-Down Approach," *Physical Review D 73* (2006): 123527.

30. Amanda Gefter, "Mr. Hawking's Flexiverse," *New Scientist 189*, no. 2548 (2006): 28–32.

31. Roger Highfield, "Stephen Hawking Plans to See Space," *Telegraph*, January 8, 2007, http://www.telegraph.co.uk/connected/2007/01/08/nhawking08.xml (accessed January 22, 2007).

32. MacAdam, "Search Is Vital."

APPENDIXES

APPENDIX A

GENERAL RELATIVITY AND COSMOLOGY

According to Einstein's general theory of relativity (1915), space and time are interwoven into a four-dimensional fabric known as space-time. This fabric can be stretched or bent into a variety of shapes, each of which being a particular solution to what are known as the Einstein field equations. These equations relate the shape of space-time to the amount and distribution of matter and energy present in the model. A helpful two-dimensional analogy is a rubber sheet.[1] When a heavy object, such as a bowling ball, is placed into the center of the sheet, it bends or deforms in a particular way in response to the size, shape, and mass of the ball. Similarly, the presence of the sun deforms the fabric of the solar system, with the resulting shape determining the orbit of the planets.

Einstein applied his equations to the universe at large, starting with two basic assumptions:

1) Matter and energy are distributed in an even manner, which means that the universe looks the same at all locations and in all directions (it is homogenous and isotropic).
2) The universe is infinitely old and in complete equilibrium. It is therefore static in nature.

To his surprise, Einstein found that his equations predicted that the universe was non-static, either expanding or contracting, so he added a "fudge factor" called the cosmological constant in order to achieve the equilibrium he expected the universe to have. His model universe closed in on itself into a sphere, like the surface of the earth (but in a higher dimension), and was finite (limited) in size but unbounded (had no edge). Dutch astronomer Willem de Sitter proposed another solution to the Einstein equations: a universe which also had a cosmological constant, but was devoid of any matter and yet was still curved. Einstein resisted accepting the solution, because he intuitively felt it was impossible to curve space-time without matter, but he eventually came to accept the idea as possible, but probably not physically relevant. Because it contained no matter (or energy, except for the cosmological constant, which may represent the inherent vacuum energy of space-time itself), a practical application for the de Sitter solution awaited the derivation of the inflationary model in the 1980s.

In 1922, Russian mathematician Alexander Friedmann published solutions without a cosmological constant that were non-static, or changing with time, including one which expanded with time, and another which was periodic in time, expanding for a while and then contracting. After the discovery of Hubble's law and the expansion of the universe, Einstein called the cosmological constant his biggest blunder and it was largely ignored. Soon after, Belgian abbé Georges Lemaître developed a model which served as the foundation of the big bang. Rather than merely having a mathematical solution to the field equations, Lemaître proposed a physically relevant model—that the universe had a distinct beginning (some finite length of time in the past) and has been expanding ever since. In 1935, H. P. Robertson and A. G. Walker independently generalized Friedmann's work to create a model known as the Friedmann-Robertson-Walker metric. The central equation related the evolution of the universe in size (including, for example, the current expansion shown in Hubble's law) to the density of matter and energy in the universe, and the curvature of the universe (and may even be generalized to include a cosmological constant). In this model, the universe has three overall (global) curvatures or geometries:

1) Flat or Euclidean[2]—The universe can be pictured (in two dimensions) as the surface of a flat rubber sheet. Two lines that begin parallel always remain parallel no matter how far they travel. The sum of the interior angles of a triangle equals 180 degrees. This geometry can be thought of as having zero curvature.

2) Spherical or Closed—This is similar to Einstein's original cosmology, in that the universe is pictured (in two dimensions) as the surface of a sphere. Lines that begin parallel will eventually merge (just as lines of longitude on the earth are parallel at the equator but meet at the poles). A triangle drawn on a sphere is distorted, such that the sum of the interior angles is greater than 180 degrees. This geometry has a positive curvature (as it curves in on itself).
3) Hyperbolical (Saddle) or Open—In this case the universe flares out like the center of a saddle, or a Pringles™ potato chip. Lines that begin parallel flare out as well, and a triangle drawn in the center of the saddle is "pinched," so the sum of the angles is less than 180 degrees. This geometry is said to have a negative curvature (the opposite of curving in on itself, or the opposite of a sphere).

The three geometries correspond to three different fates for the universe, depending on the precarious balance between gravity (which is an attractive force) and the repulsive pressure of the big bang. If the density of matter in the universe is below some critical value, nothing can counteract the initial push of the big bang, and the universe will expand forever, although slowing its rate of expansion somewhat as time goes on. If the density of matter is above the same critical value, then the collective gravity of the universe will counteract the expansion, and the universe will eventually recollapse in a big

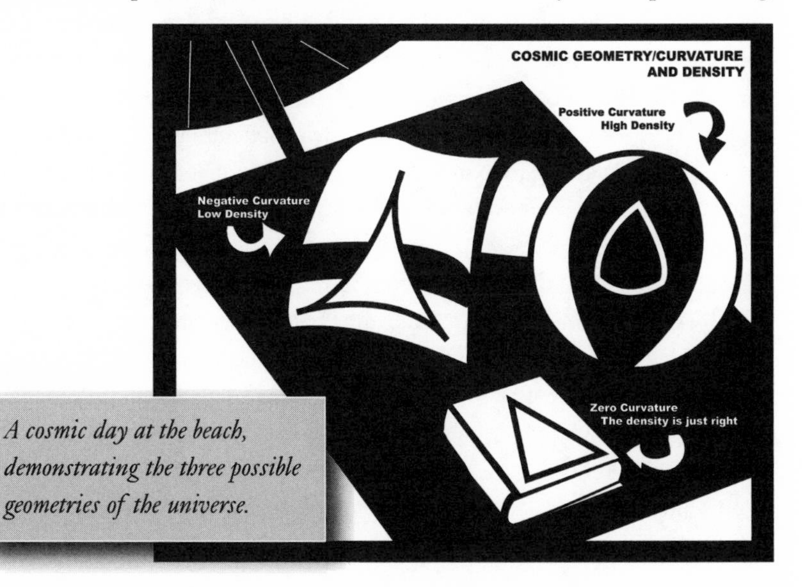

A cosmic day at the beach, demonstrating the three possible geometries of the universe.

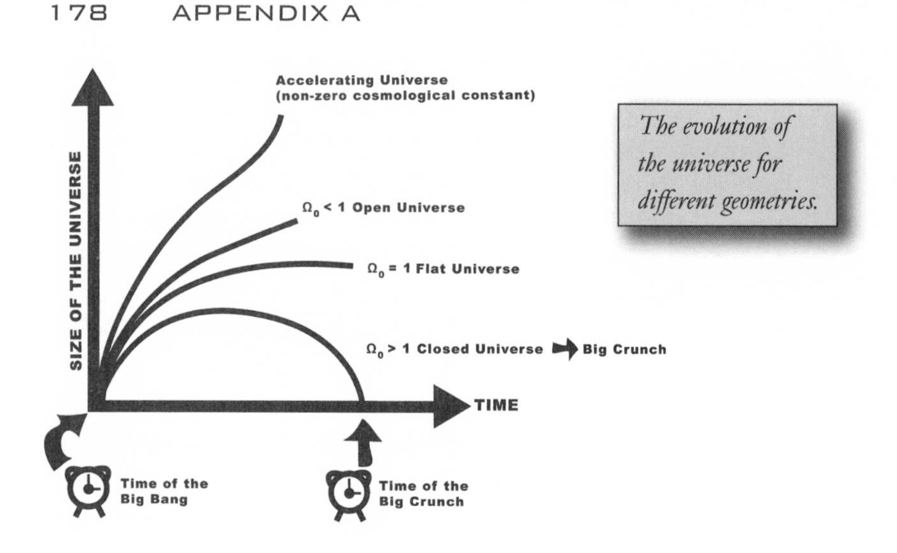

The evolution of the universe for different geometries.

crunch. There is a possibility it could then re-expand in another big bang. If the density of matter in the universe was actually the critical value, then gravity and the big bang would be balanced, and the expansion of the universe would slow to a crawl over many billions of years, but never actually stop. Thus, it is like the old story of Goldilocks and the Three Bears—one universe is too light (open), one is too heavy (closed), and one is just right (flat).

The special "just right" value for the density of matter in the universe is called the critical density, and cosmologists usually discuss it in terms of a ratio of the current density of matter (and energy) to this critical value. Therefore, in a flat universe the actual density is equal to the critical density, so the ratio of the two (called Ω—omega, the last letter of the Greek alphabet) is exactly 1. In a closed universe, the actual density is greater than the critical density, so Ω is greater than 1, and in the open universe the actual density is less than the critical density, so Ω is less than 1. Therefore, if we could observationally determine the density of matter and energy in the universe, we would theoretically know its fate.

Current observations have determined absolutely that $0.1 < \Omega < 2$, which is not a very helpful observation, because it still allows all three cases. Over the past few decades, theorists have predicted that $\Omega = 1$ (a result demanded by most models of inflation) while observations have found Ω slowly creeping up to the same special value. The most detailed observations of the cosmic background radiation done by the Wilkinson Mapping Anisotropy Probe (WMAP) give $\Omega = 1.0$ to within the uncertainties of the experiment.[3]

What is the exact breakdown of that measurement? Normal matter (so-called baryonic matter made of protons and neutrons) only accounts for 4 percent of that value. This value is sharply constrained by the nuclear chemistry of the universe. If the amount of baryonic matter were significantly higher, the universe would not have the recipe of roughly 75 percent hydrogen and 25 percent helium that astronomers currently observe.[4] Another 23 percent is made of dark matter, mysterious material that is not directly seen, but whose presence is felt through its gravitational effect on stars and galaxies. It is believed by most astronomers (and corroborated by WMAP) that most, if not all, of the dark matter is of a variety called cold dark matter. Cold dark matter (CDM) is made of particles which have mass and travel significantly slower than the speed of light. This allows them to clump together and act as "seeds" for structure in the universe (such as galaxies, galaxy clusters, and superclusters). The most popular candidates for CDM are the particles predicted by supersymmetry, known as supersymmetric partners. In particular, the neutralino, the lightest electrically neutral supersymmetric partner (yet still as much as 5,000 times more massive than a proton!), is thought to be an especially promising candidate.

This still leaves the majority of the universe—73 percent—undetermined. This is usually given the peculiar name dark energy (in symmetry with dark matter). The exact nature of this dark energy eludes scientists, but there are several tantalizing suggestions. One is that it is the inherent energy contained in the fabric of space-time—the so-called vacuum energy, whose existence is demonstrated by the Casimir effect mentioned in Chapter 5. The idea of space-time having an inherent energy density is also found in the inflationary model. There the energy density of the "false vacuum"[5] generates a temporary effective cosmological constant that exponentially inflates the universe. In such a way, Einstein's blunder has been resurrected by some cosmologists, but for a different reason. Rather than prevent the universe from evolving, the current usage of the cosmological constant is to explain how it may have evolved in its earliest moments (and how it might evolve in the future). If the cosmological constant is large and positive (greater than zero), it will counteract the tendency of gravity to pull the universe together, and the universe will expand faster and faster (as in inflation). If the cosmological constant is negative in value (less than zero), the converse is true—the universe will be doomed to collapse in on itself. Therefore, a positive cosmological constant acts like a repulsion, and a negative cosmological constant as an attraction. In the de Sitter solution, the cosmological constant is positive (as in the inflationary model). There also exists an anti-de Sitter solution, in which the cosmological constant is negative.

The existence of a non-zero cosmological constant therefore complicates the relationship between geometry and the fate previously described. It is possible for the universe to be spatially open (hyperbolic) yet doomed to recollapse due to a large, negative cosmological constant. On the other hand, it is also possible for a spherically curved universe to expand forever in the presence of a large, positive cosmological constant, or to even be static in a fine-tuned state (as Einstein had originally proposed). For example, the de Sitter solution admits positive, negative, and zero (flat) curvatures and expands forever, although the specific application of the de Sitter solution to inflation results in a spatially flat universe. By comparison, the anti-de Sitter solution has only one possible curvature—negative (hyperbolic). The natural question is, therefore, what observational evidence is there for a significant, non-zero cosmological constant, other than a possible explanation for dark energy? The proof came nearly a decade before the WMAP measurements.

Two experimental teams, the Supernova Cosmology Project (Berkeley) and the High-z Supernova Team (Mount Stromlo and Siding Spring Observatories), independently made detailed measurements of the rate of expansion of the universe using Type Ia supernovas. These brilliant outbursts are caused by the violent explosions of white dwarfs, stellar corpses the size of the earth and roughly the mass of the sun. They can be seen in distant galaxies, and their uniformity allows their distance to be accurately calculated from their apparent brightness—the dimmer the peak brightness of the supernova, the farther away it must be. What the two groups found as they probed deeper and deeper into the universe (and hence further and further back into time) is that, in direct contradiction to expectations, the expansion rate of the universe has not been slowing with time, but has actually been accelerating (speeding up) for the past 4–6 billion years. This acceleration is possibly due to a cosmological constant, or it might be due to some kind of unknown, exotic particles, giving rise to a changing force dubbed quintessence. The answer to that question awaits further observations and theoretical breakthroughs.

NOTES

1. The bowling ball/rubber sheet model represents the deformation of two-dimensional space into a third dimension at a specific moment in time. It is impossible to make a simple model of a three dimensional space deforming into a higher spatial dimension. Therefore we always use such two-dimensional models and remind ourselves that despite the vivid visuals the models provide, they cannot be taken too literally.

2. Euclidean here means that the normal laws of Euclidean geometry hold. This has nothing to do with the "trick" of using imaginary time in calculations (also referred to as Euclidean) which Hawking utilized in his work since the late 1970s.

3. A list of cosmological parameters determined by WMAP can be found at http://lambda.gsfc.nasa.gov/product/map/wmap_parameters.cfm (accessed September 23, 2004).

4. The 75/25 percentages are by relative mass. If we count numbers of atomic nuclei, the percentages are closer to 90/10, since hydrogen is lighter than helium.

5. See Appendix C for a detailed discussion of inflation.

APPENDIX B

THE LAWS OF THERMODYNAMICS AND BLACK HOLES

In Chapters 4 and 5, the laws of thermodynamics were discussed in relation to the behavior of black holes. There are four main laws, named the zeroth, first, second, and third. The zeroth law states that if two systems are in thermal equilibrium with a third system, then they are in thermal equilibrium with each other. For example, if a soda can, an ice cube, and the walls of the cooler they are contained in are all at the same constant temperature, they are in thermal equilibrium with each other, and no heat will flow between them. In black hole theory, the surface gravity of the event horizon is analogous to temperature, so the zeroth law of black hole mechanics implies that the surface gravity is constant over the entire event horizon.

The first law of thermodynamics is a statement of conservation of energy. The change in the internal energy of a system is equal to the heat added to the system minus the work that had to be done by the system. Energy can be converted from one form to another (the friction generated by the brakes of a car into heat, for example), but it is ultimately conserved. The analogous law for black holes is that a change in the mass of the black hole is related to a change in the area of the event horizon.

Several different statements of the second law of thermodynamics are commonly found. The most intuitive is that heat naturally flows from a hot object to a cold one. The heat applied to the bottom of a frying pan naturally flows into the cold food contained inside, cooking the food. The converse is not true, however. Heat will not flow from a cold object to a hot one. Since heat and temperature are related to a quantity called entropy (a measure of the disorder of the system), the second law is sometimes stated that in a closed system, the entropy always increases. Since the universe is, as we define it, a closed system, the entropy of the universe increases as it evolves. The analogous law for black holes predicts that the surface area of the event horizon always increases. If two black holes coalesce, then the total surface area of the new black hole must be larger than that of the two "ingredient" black holes.

The lowest possible temperature in nature is called absolute zero, 273 degrees below zero Celsius. The third law of thermodynamics has two forms, the weaker (Nernst) form, and stronger (Planck) form. The weaker form states that it is impossible to lower the temperature of a system to absolute zero in a finite number of steps. This is analogous to the third law of black hole mechanics, which states that it is not possible to reduce the surface gravity of a black hole to zero in a finite number of steps. The stronger form of the third law states that entropy of a system goes to zero as its temperature does the same. The third law of black hole mechanics does not, in and of itself, prevent the entropy of a black hole from reaching zero; however, if the entropy (and hence surface area) of a black hole *were* to reach zero it would obviously create a naked singularity.

APPENDIX C

INFLATIONARY COSMOLOGY

PROBLEMS WITH THE "CLASSIC" BIG BANG MODEL

A common misconception about the cosmological inflation paradigm is that it replaces the big bang. It is actually a modification of the classic big bang, developed to answer several problems and nagging questions. Although the exponential expansion which gives the theory its name only lasted for an infinitesimally small fraction of a second, its effects are visible for the entire history of the universe. Alan Guth, the "father of inflation," explains that it "supplies the beginning to which the standard big bang theory is the continuation."[1]

Different authors list varying numbers of problems with the big bang theory, which inflation purports to solve, but the four key examples will be discussed here:

1) The magnetic monopole problem.
2) The smoothness problem.
3) The flatness problem.
4) The density perturbation problem.

1) Grand unified theories, which seek to describe the combination of the strong, weak, and electromagnetic forces in the high temperatures of the

first 10–35 seconds of the universe's existence, predict there should exist massive particles named magnetic monopoles. These can be thought of as isolated north or south magnetic poles. The problem is the high number of these as yet unseen particles predicted to exist—a trillion times more (by mass) than all the observed matter in the universe! Increasing the density of the universe by such a mammoth amount would not only make it closed geometrically (dooming it to a fiery "big crunch"), but it would have slowed the expansion of the universe to the rate currently observed in about 30,000 years. Given that the age of the universe is now estimated to be between 13 and 14 billion years, there is clearly a serious problem between theory and observation.[2]

2) Measurements of the cosmic background radiation, the echo, or energy fingerprint, of the big bang, have consistently shown that it is extraordinarily smooth, to a few parts in 100,000. This is difficult to explain if one considers what this means— either the initial big bang was absolutely uniform in temperature to this amount (i.e., it was created in remarkable thermal equilibrium) or somehow the different parts of the expanding universe were able to communicate with each other far faster than the speed of light and reach equilibrium at some point afterwards. This is also called the horizon problem, for the following reason. As you look up into the night sky, you could (with a telescope of the greatest theoretical power), see as far as 13.5 billion light years in each direction, if the universe is 13.5 billion years old. This is because you can only see light which has had sufficient time to reach your eye. A year from now, you could see 13.5 billion + 1 light years in each direction, and so on. This means that observable patches of the universe on opposite parts of your sky are separated by 27 billion light years, or it would take 27 billion years for a light signal to travel from one side of your horizon to the other. How, then, can the cosmic background radiation on opposite sides of the sky be at the same temperature to within a few parts in 100,000? One might think it would be possible for information to have been freely exchanged when the universe was much smaller, but it was initially expanding so fast that it contains at least 1083 causally disconnected regions that have still not had sufficient time to communicate.[3]

3) The Friedmann-Robertson-Walker model described in Appendix A admits three possible geometries or curvatures for the overall uni-

verse—open (hyperbolic), closed (spherical), or flat. Which curvature the actual universe has, depends on the density of matter and energy it possesses relative to a critical value (referred to in scientific shorthand by Ω). If $\Omega > 1$, the universe is "too dense" and will recollapse. If $\Omega < 1$, there is insufficient density to close the universe, and it will expand forever at an ever-slowing rate. The case of $\Omega = 1$, a flat universe, is a finely tuned value, often compared to balancing a pencil on its point. It is possible that the universe was created at exactly this special value, but there is no reason in the classic big bang model why it should have chosen such an unstable value. The reason it is unstable can be seen from the pencil analogy—any slight deviation or fluctuation to one side will send the pencil toppling over. Likewise, any perturbation from the absolute value $\Omega = 1$ will mushroom over time, driving Ω farther away from 1 as the universe evolves. In order for Ω to be as close to one as we observe today, it would have originally had to have been precisely equal to 1, to 1 part in 10^{49} or more.

4) Despite the remarkable smoothness of the cosmic background radiation, it is not completely smooth, otherwise, we would not exist today. The minute fluctuations in temperature now observed are a reflection of the fact that matter is not smoothly distributed in the universe. Instead there exist clumps of matter, in the form of galaxies, clusters of galaxies, and superclusters. There must have existed "seeds" in the early universe from which these large scale structures grew—fluctuations or perturbations in the density of matter. The original big bang model is silent on the source of these all-important deviations from perfect homogeneity.

SYMMETRY BREAKING AND THE HIGGS PARTICLE

Inflation relies on the concept of fields. A field is some physical quantity that can be defined over a region of space-time. The best-known examples of fields are electric fields and magnetic fields (sometimes spoken of under the generic title of electromagnetic fields), and the gravitational field. Fields are invisible, but they affect matter and thus can be measured and studied. For example, a charged particle such as an electron will move in a particular way in a magnetic field. The interactions of fields (with each other and with matter) are governed by field equations, such as Einstein's field equations of general relativity. In quantum mechanics, every field can be thought of as being represented by a particle, and vice versa (the so-called wave/particle duality). For example, the electromagnetic field is connected with the photon, a particle of light.

The simplest kind of field is a scalar field, in which only the value of the field at a point, not its direction, is measured. For example, a weather map that plots temperatures represents a scalar field, whereas a weather map that plots wind speed and direction does not. Scalar fields are represented by spinless (spin 0) particles, the most well-known example of which is the Higgs particle, proposed by Peter Higgs in 1964 to explain why particles such as electrons have mass. The Higgs particle explains what is called a "broken symmetry" in physics. An everyday example of symmetry breaking can be found at a round banquet table. Typically the eight or so guests will sit down, look around, and be uncertain as to which bread plate or drinking glass is supposed to be theirs. There is a therefore perfect symmetry of the dishware on the table. However, after an awkward moment or two, one bold person will reach out and grab a glass, or subtly drag one of the bread plates closer to their main plate, thus breaking the symmetry for the entire table. If the person chose the glass to their right, so must everyone else at the table. In quantum field theory, symmetry-preserving states are unstable, as if the symmetry wanted to be broken. The Higgs particle (and its associated field) provides the mechanism by which the symmetry is broken.

In grand unified theories there are multiple Higgs fields, the actual number varying from model to model. The symmetry of the model is preserved when all Higgs fields are zero, but as soon as at least one of them takes a value other than zero, the symmetry is broken. It is simplest to picture a toy model where there are two Higgs fields, which can be represented pictorially like a two-dimensional grid (like a checkerboard). The center of the grid represents both Higgs fields having a value of zero. A third dimension—height—can be added to our picture, representing the energy density of space-time. The lowest possible value of this energy density (the "bottom" of the 3-d graph) is called the true vacuum. If this is called the true vacuum, there must be a corresponding false vacuum. Picture a Mexican sombrero with a slight dimple or dent in the very top of the hat. If we look just at the center of the hat, the bottom of the dimple looks like the lowest point of the hat. It represents a local minimum in the height of the hat. However, a wider view shows that this is not the true lowest point in the hat—the round trough in the brim of the hat has that honor. This is the global minimum of the hat. The dent in the top of the hat is the false minimum or false vacuum, whereas the trough in the brim is the true minimum or true vacuum. Note that the dent at the center is a state of symmetry, as all the Higgs fields are zero, whereas the brim of the hat is a state of broken symmetry, as the Higgs fields are not zero. The surface of the hat represents how the energy density depends on the values of the Higgs fields.

Now consider a small ball rolling on the surface of the hat. The act of rolling from the top to the bottom represents a phase transition, from a state of preserved symmetry to a state of broken symmetry. It is possible that the ball could get stuck in the false vacuum in the top of the hat and be unable to complete the transition to the broken symmetry state. But recall that nature wants to break the symmetry, so there is a price to be paid for being caught in the false vacuum. If the ball were a bubble of the big bang, it would be in a supercooled state (like freezing rain remaining liquid below 32° F). While stuck in the false vacuum, the bubble would be under the influence of a repulsive force, and it would inflate exponentially, growing 10^{25} times bigger in the blink of an eye. Such an exponentially expanding universe is described by the de Sitter solution to Einstein's field equations, as discussed in Appendix A. The universe inflates so much that the density of matter declines to essentially zero, making the matterless de Sitter model a good approximation. However, we obviously do not currently live in a de Sitter universe, so the bubble must somehow get out of the false vacuum state and reach the broken symmetry state of the true vacuum.

OLD INFLATION VS. NEW INFLATION

In Guth's original inflationary model, the bubbles cannot get out of the false vacuum "dimple" by rolling up and over the hump, because they don't have sufficient energy to do so. Instead, they tunnel out through the hump. This is clearly impossible by the classical laws of physics, but it is common enough in the strange world of quantum mechanics. Quantum tunneling is the basis of such technological advances as the tunnel diode in electronics and the scanning tunneling microscope. There is also a derivation of Hawking radiation that has particles tunneling out past the event horizon from the interior of the black hole. In this so-called old inflation, the universe tunneled out one bubble at a time, and somehow the bubbles collided and recombined on the other

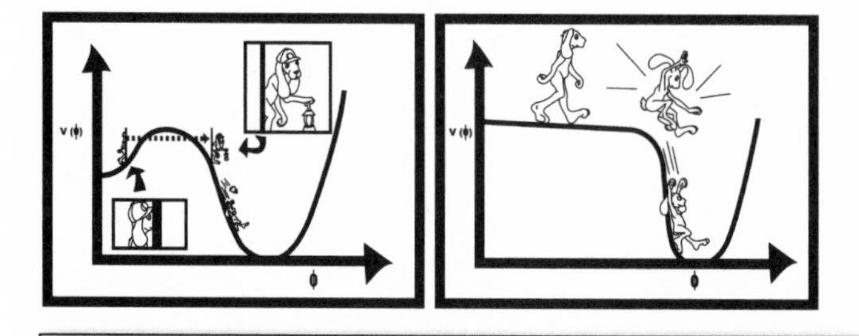

The evolution of bubbles of false vacuum in old inflation (left) and new inflation (right).

side—in the broken symmetry state. However, as Hawking and others showed, the density perturbations were much too large to result in the universe we see today. Thus old inflation suffered from what is called the graceful exit problem—there is no realistic way to end the inflationary phase.

The new inflation model of Albrecht/Steinhardt and Linde avoided this problem by not having bubbles try and reconnect. Instead, the entire visible universe was spawned from a single bubble of false vacuum. The mechanism for leaving the false vacuum state was different as well. Instead of an energy barrier, which had to be tunneled through, the shape of the hat included a long, flat region, giving this model its nickname of the "slow rollover." Quantum fluctuations would start the ball rolling, in a literal sense, slowly driving the bubble off of the false vacuum plateau, inflating as it rolled. When the ball reached the bottom of the hill (the brim of the sombrero), it would oscillate back and forth, reheating the supercooled universe to the level expected by the big bang model. Some of the energy would be converted into particles, repopulating the now-empty universe.

Inflation was developed in response to problems with the original big bang model. How successful was it in solving those problems? At the end of the inflationary period, the density of any particles created beforehand was diluted to essentially zero by the huge increase in the volume of the universe, thus solving the magnetic monopole problem. This also explains why, as of yet, there has been no experimental observation of magnetic monopoles—the nearest one could now be light years away!

The incredible expansion of the universe also solves the smoothness problem in much the same way. Any inhomogeneities in the background radiation that existed prior to inflation would be smoothed out, just as wrinkles in fabric (or skin) disappear when it is stretched. Another way of looking at the problem is that it is no longer assumed that the entire visible universe arose from a state of thermal equilibrium, only that what we see was born from a tiny patch of the initial big bang that, in itself, had achieved a constant temperature over its small volume. A solution to the flatness problem follows close behind. Any large-scale curvature to space-time that was created by the initial big bang would be flattened out, as the surface of a sphere looks flat when it becomes large enough (like the Earth).

The density perturbation problem was the most complicated to solve, and it can be argued that inflation only gives a provisional solution until more precise measurements of the cosmic background radiation can be made. The exponential expansion of the universe ends when the phase transition is made from false vacuum to true vacuum, or, in pictorial terms, the scalar field ball reaches the steep part of the energy hill. The attendees of the Nuffield Workshop in 1982 realized that quantum fluctuations would affect the rolling along the plateau, and that these fluctuations would not be the same everywhere. As Alan Guth summarized, "The nonuniformities in the rolling develop into perturbations in the mass density of the universe, which might be the primordial seeds from which the present structure of the universe emerged."[4]

Over the past two decades, a number of different inflationary models have been proposed, some in the usual four-dimensions of our universe, and some in higher dimensions (such as brane models), several of which were proposed by Hawking and his collaborators. The shape of the energy curve (the hat) varies significantly from model to model, as does the field responsible for inflation. No longer is the Higgs particle the "only game in town." All models try, in their own way, to answer the lingering questions about inflation—how does it begin, how does it end, and what field or force drives it in-between. Density perturbations continue to be an important issue, as any model must be fine-tuned to give perturbations of a size and pattern consistent with current observations. As observations of the cosmic background radiation improve in their precision and ability to discern tiny deviations, some models fall by the wayside while others remain in the running. The WMAP has observed such a high level of detail that it "both confirms the basic tenets of the inflationary paradigm and begins to quantitatively test inflationary models."[5] The big bang seems permanently enhanced by the idea of inflation, yet there is still plenty of room for future researchers to fill in the missing pieces in our understanding.

NOTES

1. Tom Yulsman, "Give Peas a Chance," *Astronomy* (September 1999): 41.

2. Alan H. Guth, "Inflation," *Proceedings of the National Academy of Sciences* 90, no.11 (1993): 4874.

3. Alan H. Guth, "Inflationary Universe: A Possible Solution to the Horizon and Flatness Problems," *Physical Review D* 23 (1981): 347.

4. Guth, "Inflation": 4874.

5. H. V. Peiris et al., "First Year Wilkinson Microwave Anisotropy Probe (WMAP) Observations: Implications for Inflation," *astro-ph*/0302225 (2003): 35.

APPENDIX D

THE AdS/CFT CORRESPONDENCE

In the process of trying to extend string theory to black holes (and search for an answer to Hawking's black hole information loss paradox), Gerard 't Hooft developed what is called the holographic principle.[1] This principle asserts that for some region of space, all the information contained inside it can be represented by the region's boundary. This is similar to a hologram, a two-dimensional projection of a three-dimensional object made with lasers. Applied to black holes, this means that the information contained within the event horizon (such as the identity of objects which fell in) should be somehow encoded by the event horizon.

In brane theory, our universe is depicted as a four-dimensional membrane existing in a fifth dimension or bulk. Other dimensions may exist as well. In some sense, our universe is merely a shadow being projected by the bulk dimension. Our universe is the boundary and the bulk is the region. Therefore, studying physics on the boundary (in our universe) tells us about the physics of the higher dimension, and vice versa. It is as if we are shadow puppets, and by studying the details of the shadows we can figure out the way the hands are arranged in higher dimensions.

One of the most important special cases of the holographic principle is the AdS/CFT correspondence or duality. It is difficult to comprehend on a nontechnical level, but dissecting the name may help. Also named the Maldacena conjecture, after its Argentine creator, Juan Maldacena,[2] the acronym stands for anti-de Sitter/Conformal Field Theory (AdS/CFT). Anti-de Sitter space is a variation of the de Sitter space used in inflationary models and can be roughly pictured as "a universe in a cavity. The walls of the cavity behave like reflecting surfaces so that nothing escapes it."[3] It has a relatively simple geometry, which means that its properties have been extensively studied, and even though it is not an especially realistic model of the observed universe, calculations done using this space are more simple than in other geometries. It differs from de Sitter space in several ways; the most basic is in having a negative cosmological constant, rather than a positive one. The basic differences between de Sitter and anti-de Sitter space are discussed in Appendix A. A conformal field theory is a theory of fields in which the equations have a particular mathematical symmetry (i.e., the theory does not vary under certain mathematical transformations, like a rotation). For example, electromagnetism has this particular symmetry. A duality connects two apparently different physical theories and shows that they are equivalent. This is extremely useful when one theory is easier to calculate than the other.

The AdS/CFT correspondence states that quantum gravity, (e.g., a superstring or supergravity theory) defined in the five-dimensional anti-de Sitter space, is equivalent to another quantum field theory living on the four-dimensional boundary of the AdS space. Hawking's papers on brane worlds and brane black holes made extensive use of the AdS/CFT correspondence, as did his apparent solution of the black hole information loss paradox. As complex as this mathematical technique appears, it actually allows for the simplification of much more complex calculations, allowing for the properties of selected models of black holes and cosmology to be studied in some detail.

NOTES

1. G. 't Hooft, "Dimensional Reduction in Quantum Gravity," in *Salamfestschrift: a Collection of Talks*, ed. A. Ali. J. Ellis and S. Randjbar-Daemi (Singapore: World Scientific, 1993).

2. J. Maldacena, "The Large N Limit of Superconformal Field Theories and Supergravity," *Advances in Theoretical Mathematical Physics* 2 (1998): 231–52.

3. Leonard Susskind, "Superstrings," *Physics World* (November 2003): 34.

GLOSSARY

AdS/CFT conjecture • The principle that a quantum field theory in five-dimensional anti-de Sitter space can be modeled by another field existing on its four-dimensional boundary.

Amyotrophic lateral sclerosis (ALS) • An unexplained, incurable, debilitating disease marked by the deterioration of the body's ability to control voluntary muscular functions (such as motion).

Anthropic principle • The proposal that the universe is the way it is because if it were different we would not exist.

Anti-de Sitter space • A solution to Einstein's field equations of general relativity that describes a mathematically simple but physically unrealistic model of the universe.

Baby universe • In imaginary time, an independent, closed universe connected to our universe by a wormhole.

Baryonic matter • Normal matter made of protons and neutrons.

Big bang model • A scientific theory concerned with the creation of the universe in a huge "explosion" approximately fourteen billion years ago.

Black hole information paradox • The idea that there is a permanent loss of information about the properties of objects that fall into a black hole, in violation of the principles of quantum mechanics.

Boson • A particle whose spin is a whole number and does not obey the Pauli exclusion principle.

Boundary conditions • The properties of a system that define its beginning state of being, or more generally, any boundary of the system in space or time.

Brane • In M-theory, a fundamental object specified by its number of dimensions in space (for example, one dimension denoting a string, two a membrane).

Casimir effect • The observed attraction between two conducting metal plates in a vacuum, evidence for the existence of virtual particles.

Chronology protection conjecture • Hawking's proposal that the laws of physics do not allow time travel by objects above the atomic level.

Closed time-like curve • Loops in time that act like a time machine, allowing a particle to go backward in time.

Closed universe • A universe that will eventually collapse in on itself because the gravitational pull of all the matter and energy contained within is greater than the outward force of the big bang.

Cold dark matter • Dark matter that moves much slower than the speed of light and is able to clump together and possibly form the "seeds" for structure in the universe.

Compact metrics • Self-contained space-times that bend back on themselves and therefore have a finite size (such as the surface of a sphere).

Conformal field theory • A field theory whose equations show a particular symmetry, meaning it remains the same when certain mathematical operations are performed on it.

Cosmic background radiation • Pervasive leftover energy from the big bang, which cools over time and currently has a temperature of 2.7 K.

Cosmic censorship conjecture • Penrose's proposal that the singularities in black holes are always hidden behind an event horizon.

Cosmological arrow of time • The flow of time onward from the beginning of the universe.

Cosmological constant • Einstein's addition to his equations of general relativity in order to make the universe static.

Cosmological horizon • The greatest distance one can theoretically see in any direction in space, determined by the speed of light and the age of the universe.

Cosmology • The scientific study of the structure, composition, and evolution of the universe.

Critical density • The density of a flat universe, which balances the inward pull of gravity and the outward push of the big bang.

Dark energy • A mysterious energy that comprises 73 percent of the universe and is responsible for the current acceleration of its expansion.

Dark matter • Unseen material whose presence is only known by the gravitational effect it has on other objects, such as stars.

de Sitter space • A solution to Einstein's field equations of general relativity, which describes an empty, exponentially expanding universe and is used to describe inflationary models.

Doppler effect • Wavelengths of light appear to lengthen (shift toward the red) as the source moves away from an observer, while the wavelengths appear to shorten (shift toward the blue) as the source approaches an observer.

Duality • A connection between two seemingly different physical theories that shows they give the same results.

Einstein field equations • The basic equations of general relativity that determine how matter and energy warp space-time.

Entropy • The disorder of a physical system, which is related to the number of possible microscopic configurations that system might have.

Euclidean approach • Quantum gravity calculations that utilize imaginary time.

Euclidean space-time • A space-time that includes imaginary time.

Event horizon • The boundary of a black hole.

False vacuum • A local minimum value of the energy of a system that is not the overall lowest energy value.

Fermion • A particle whose spin is a multiple of 1/2 and obeys the Pauli exclusion principle.

Field • A physical quantity that can be defined for a particular region of space-time, such as a magnetic field or a gravitational field.

Flat universe • A universe in which the force of gravity is perfectly balanced by the initial push of the big bang.

Friedmann-Robertson-Walker model • A mathematical model of the universe that can have three possible curvatures and corresponding fates: open, closed, and flat.

General theory of relativity • Einstein's explanation of gravity as the warping of space-time by the presence of matter and energy.

Grand unified theories • Scientific models that explain how the strong nuclear force, weak nuclear force, and electromagnetic forces are all related under conditions of extremely high temperature (such as in the very early universe).

Grandfather paradox • In theories of time travel, the paradox that someone might be able to go back in time and prevent their grandparents from giving birth to their parents (and therefore preventing their own birth).

Hawking radiation • Hawking's prediction that black holes have an effective temperature and radiate; the lower the black hole's mass, the higher the temperature and the greater their radiation.

Heisenberg uncertainty principle • A principle of quantum mechanics, which states that it is impossible to know with certainty both the position and momentum of a particle, or the energy of an event and its duration in time.

Hierarchy problem • The question as to why gravity is so much weaker than all the other fundamental forces of nature.

Higgs field • A quantum mechanical field (and corresponding particle) theorized to be responsible for elementary particles having masses.

Holographic principle • The principle that the properties of a volume of space-time can be found by studying the boundary of that region.

Homogeneous • Appears the same at all places.

Hubble's law • The discovery in the 1920s that the farther away a galaxy is, the faster it appears to be receding from us; interpreted as evidence of the expansion of the universe and the big bang.

Imaginary numbers • A mathematical technique developed in order to define the square root of negative numbers.

Imaginary time • Time measured in imaginary numbers.

Inflationary cosmology • The proposal that the early universe underwent a period of exponential expansion.

Irreducible mass • A specific combination of a black hole's mass and rotation, which can only increase as a black hole interacts with the universe.

Isotropic • Appears the same in all directions.

Lou Gehrig's disease • An American term for ALS.

Magnetic monopole • A hypothetical particle that represents an isolated north or south magnetic pole.

Motor neurone disease • A British name for ALS.

M-theory • A higher-dimensional theory that includes string theory and supergravity as special cases and whose basic properties are not currently understood.

Naked singularity • A singularity not hidden inside an event horizon.

New inflation • A revision of the inflationary cosmology in which the entire visible universe evolved from a single bubble of the big bang.

No-boundary proposal • Hawking's proposal that, when measured in imaginary time, the universe has no edge or boundary.

Open universe • A universe that will expand forever because the gravitational pull of all the matter and energy contained within is less than the outward force of the big bang.

Particle physics • The scientific study of the basic building blocks of nature, such as electrons and quarks.

Pauli exclusion principle • In quantum mechanics, the concept that no two identical fermions (e.g., no two electrons) can occupy the same quantum state at the same time.

Phase transition • A change between two physical states, such as water freezing from liquid to solid.

Primordial black holes • Tiny black holes proposed by Hawking to have been created in the very early universe.

Psychological arrow of time • The everyday experience of the passage of time in the forward (future) direction.

Quantum gravity • Any scientific model that attempts to unite quantum mechanics and general relativity.

Quantum mechanics • Scientific laws that govern the microscopic interactions of fundamental particles and atoms.

Quantum tunneling • In quantum mechanics, the ability of a particle to move from one side of a barrier to another without having enough energy to move over the barrier.

Scalar field • A field whose only important property is that it has different values at different points in space-time.

Schwarzschild radius • The size of the event horizon of a simple (nonrotating, electrically neutral) black hole.

Singularity • A point where the curvature of space-time is so extreme that general relativity breaks down; the big bang and the center of a black hole are the best-known examples.

Singularity theorems • Mathematical proofs that determine the physical conditions under which a singularity forms, such as in the creation of a black hole.

Space-time • The four-dimensional fabric of the universe, which interweaves three dimensions of space and one of time.

Special theory of relativity • Einstein's explanation of how the laws of physics appear the same to all observers moving at a constant speed in relation to each other.

Spectroscopic binary • A system in which the presence of two stars is surmised because of the periodic Doppler shift of the light emitted by the stars as they orbit each other.

Standard model • The summation of the basic principles of grand unified theories and particle physics.

Steady state model • A scientific model for the universe as an eternal object that was rejected by observational evidence in the late 1960s.

String theory • A scientific theory that envisions elementary particles as different modes of vibration of tiny one-dimensional objects.

Strong nuclear force • A fundamental force of nature that governs how quarks combine to form protons and neutrons.

Sum over histories method • A mathematical technique developed by Richard Feynman that analyzes an event in quantum mechanics by examining the various ways a particle could have gotten from the beginning to the end, and summing up their relative contributions.

Supercooling • The process of cooling a liquid below its freezing temperature without it turning into a solid; in cosmology the term is used to describe the process of cooling a region of the early universe below some critical temperature without some fundamental force of nature freezing out and becoming separate from the others.

Supergravity • A scientific theory that combines supersymmetry and general relativity.

Superstring • A string that gives rise to both bosons and fermions.

Supersymmetry • A scientific theory that explains the relationship between bosons and fermions.

Symmetry breaking • A physical system that moves from a symmetric state to a nonsymmetric one.

Theory of everything • A scientific theory that attempts to achieve the unification of all four fundamental forces and describes the conditions of the very early universe.

Thermodynamic arrow of time • The fact that entropy increases with time.

Thermodynamics • The laws governing the statistical behavior of large systems of atoms, including heat and energy transfer.

True vacuum • The lowest possible value of the energy of a system.

Uniqueness theorems • Mathematical proofs of the limited number of distinguishing factors a black hole may possess (mass, angular momentum, and electric charge).

Vacuum energy • The inherent energy contained within the fabric of space-time.

Virtual pair • Under the provisions of the Heisenberg uncertainty principle, a particle and its antiparticle can be temporarily created out of the vacuum but must annihilate within a short period of time in order for the total energy to be conserved.

Weak nuclear force • A fundamental force of nature that governs radioactive decay.

Wormhole • A tunnel that connects two different regions of space-time in one universe, or two universes to each other; theoretically they could be used as a shortcut to travel in space or to travel backward in time.

SELECT BIBLIOGRAPHY

Baez, John. "Week 207." *This Week's Finds in Mathematical Physics*, July 25, 2004. http://math.ucr.edu/home/baez/week207.html.

Banks, Thomas. "The Cosmological Constant." *Physics Today* (March 2004): 46–51.

Bekenstein, Jacob D. "Black Hole Thermodynamics." *Physics Today* (January 1980): 24–31.

Boslough, John. *Stephen Hawking's Universe.* New York: Quill/William Morrow, 1985.

Falk, Dan. "The Anthropic Principle's Surprising Resurgence." *Sky and Telescope* (March 2004): 43–47.

Ferguson, Kitty. *Stephen Hawking: Quest for a Theory of Everything.* New York: Franklin Watts, 1991.

Filkin, David. *Stephen Hawking's Universe.* New York: Basic Books, 1997.

Folger, Tim. "The Ultimate Vanishing." *Discover* (October 1993): 98–106.

Gibbons, G. W., E. P. S. Shellard, and S. J. Rankin, ed. *The Future of Theoretical Physics and Cosmology.* Cambridge: Cambridge University Press, 2003.

Greene, Brian. *The Elegant Universe: Superstrings, Hidden Dimensions, and the Quest for the Ultimate Theory.* New York: W. W. Norton and Co., 1999.

Guth, Alan H. *The Inflationary Universe: The Quest for a New Theory of Cosmic Origins.* New York: Perseus Books, 1997.

Hawking, Jane. *Music to Move the Stars: A Life with Stephen Hawking.* Rev. ed. London: Pan Books, 2000.

Hawking, Stephen W. "The Quantum Mechanics of Black Holes." *Scientific American* (January 1977): 34–40.

———. *A Brief History of Time.* Toronto: Bantam Books, 1988.

———. *Black Holes and Baby Universes and Other Essays.* New York: Bantam Books, 1993.

———. *Hawking on the Big Bang and Black Holes.* Singapore: World Scientific, 1993.

———. *The Universe in a Nutshell.* New York: Bantam Books, 2001.

———. *The Theory of Everything.* Beverly Hills: New Millennium Press, 2002.

———, ed. *Stephen Hawking's A Brief History of Time: A Reader's Companion.* New York: Bantam Books, 1992.

———. *Professor Stephen Hawking's Web Pages.* http://www.hawking.org.uk.

Hawking, Stephen W., and Roger Penrose, *The Nature of Space and Time.* Princeton: Princeton University Press, 1996.

Muscular Dystrophy Association. *When a Loved One Has ALS: A Caregiver's Guide.* Tucson: MDA, 2003.

Naeye, Robert. "Delving into Extra Dimensions." *Sky and Telescope* (June 2003): 38–44.

Overbye, Dennis. "The Wizard of Space and Time." *Omni* (February 1979): 44–107.

Susskind, Leonard. "Superstrings." *Physics World* (November 2003): 29–35.

Thorne, Kip S. *Black Holes and Time Warps.* New York: W. W. Norton and Co., 1994.

Yulsman, Tom. "Give Peas a Chance." *Astronomy* (September 1999): 38–46.

INDEX

Accidental Marathon, The (Hawking, Lucy), 164
AdS/CFT conjecture, 143n42, 153, 168, 193–94
Albrecht, Andreas, 93n17, 190
Alien Planet (TV show), 163
Al-Khalili, Jim, 163
Allen, Bill, 112
ALS (Amyotrophic lateral sclerosis), 10, 41, 75, 82, 119, 124, 134, 147, 152, 158–59
American Physical Society, 132
Amnesiac (Radiohead), 131
Anderson, Susan, 112
anthropic principle, 102, 103, 104, 130–31, 169–70
anti-de Sitter/Conformed Field Theory. *See* AdS/CFT conjecture
anti-de Sitter space, 137–38, 143n42, 194
  See also de Sitter space
Archon X Prize for Genomics, 166–67
arrows of time, 100–102
Astrophysical Journal, 72
atheism, 29, 43, 76, 108
  See also religion
At Home in France (Hawking, Jane), 153

Babylon 5 (TV show), 111
baby universe, 110, 111, 153
Back to the Future (film), 118
Bantam Books, 96, 107, 121, 126
Bardeen, James, 60, 63
Beckham, David, 162
Bekenstein, Jacob, 59–61, 62, 137
Berman, Robert, 27, 31, 120
Berry, Gordon, 30, 31
Beyond the Horizon (film), 163
big bang, 47, 48, 54, 55, 61–62, 84, 86, 100, 102, 121, 176, 177–78
  big bang model, 32–33, 83, 85, 93n13, 185–92
  singularity and, 54, 55
  See also cosmic background radiation; inflation
"Black Hole Explosions" (Hawking), 64
black holes, 9, 38, 46, 55–56, 59, 60–61,

64, 72–73, 77, 101, 108, 109, 110, 111, 121, 131, 136–37, 150, 154, 157, 169
baby universe and, 111, 153
branes and, 138, 139, 194
entropy and, 56, 57, 60, 76, 137, 147, 184
exploding, 61, 64, 69
horizon, 46, 56–57, 60, 63, 76, 137, 153, 183, 184, 189, 193
information paradox and, 46, 60, 71, 111, 137, 138, 152–53, 154, 168, 169, 193, 194
primordial, 55, 62, 64, 69
quantum mechanics and, 62–63, 65, 71, 75–76, 137, 183
radiating, 62, 64, 65, 66, 69–70, 71, 76, 90
singularity and, 11, 54, 55, 111
thermodynamics and, 55–56, 60–61, 63, 76, 183–84
uniqueness theorems and, 55, 60, 71
"Black Holes" (Hawking), 57
Black Holes and Baby Universes and Other Essays (Hawking), 121
Bloom County (comic strip), 112
Bohr, Neils, 64
Bondi, Hermann, 37, 40, 48, 54
Boslaugh, John, 96
"Boundary Conditions of the Universe, The" (Hawking), 84
Bousso, Raphael, 138
branes, 137–38, 139, 143n2, 146, 153, 191, 193, 194
Branson, Richard, 166
"Breakdown of Predictability in Gravitational Collapse" (Hawking), 71
Breathed, Berke, 112
briefer, 163
Briefer History of Time, A (Hawking), 163, 167
Brief History of Time, A (Hawking), 10, 108, 110, 123, 126, 134, 145, 146, 162
film version of, 119
illustrated version, 123
British Comedy Awards, 162
Bryan, Richard, 30, 31
Bulletin of Atomic Scientists (journal), 164
Burgoyne, Chris, 131
Bush, George W., 165

California Institute of Technology (Caltech), 48, 66, 69, 70–71, 73, 74, 75, 76, 115, 125, 136, 151, 167
Cambridge, 21, 30, 32, 33, 34, 37–39, 40, 42–43, 44, 46, 47, 48, 53, 62, 64, 74, 76, 78, 81, 83, 87, 95, 97, 107, 110, 111, 112, 122, 123, 129, 134, 135, 148, 149, 150, 152, 153, 170
See also Department of Mathematics and Theoretical Physics (DAMTP); Gonville and Caius College
Campaign for Nuclear Disarmament, 87
Carey, Jim, 150
Carr, Bernard, 66, 69, 73
Carter, Brandon, 47, 60, 63, 103
Casimir effect, 63, 118, 179
Catalyst (journal), 166
Catholic League, 165, 166
CDM. *See* cold dark matter
Celebrity Big House (TV show), 162
CERN, 96, 102, 150
Chandrasekhar, Subrahmanyan, 46, 73
Charles (prince), 133
Choptuik, Matthew, 125
Christodoulou, Demetrios, 56, 57, 125

chronology protection conjecture, 119
Church, Michael, 24
Clinton, Bill, 125
closed time-like loops. *See* time travel
closed universe, 90, 97, 101, 111, 129, 178
cold dark matter, 179
Coles, Peter, 140–41
Collins, C. B., 103
Communications in Mathematical Physics (journal), 65
compact metrics, 84, 88
Conseil European pour la Recherche Nucleaire. *See* CERN
Contact (Sagan), 111, 117
Copernicus, Nicholas, 62, 73, 149
cosmic background radiation, 48, 54, 90, 146, 178, 186, 187, 190–91
cosmic censorship conjecture, 57, 125
cosmological constant, 79n21, 176, 179–80, 194
cosmology, 25, 32, 37, 39–40, 44, 46–47, 100, 124, 147, 150, 194
  general relativity and, 175–81
  inflationary cosmology, 84, 85, 87, 170, 185–92
  See also big bang; inflation
COSMOS supercomputer consortium, 124, 152
critical density, 177, 178, 187
Cutler, Curt, 153
Cygnus X-1, 72–73, 115

Daily Telegraph (newspaper), 140
dark energy, 179, 180
dark matter, 179
Davies, P. C. W., 65
Dennis Sciama Memorial Lecture, 161
density perturbations, 86, 87, 90, 103, 185, 190, 191
Department of Mathematics and Theoretical Physics (DAMTP), 38, 61, 77
  See also Cambridge
de Sitter space, 75, 87, 138, 194
  See also anti-de Sitter space
Dicke, Robert, 102
Dickie, Brian, 147
Dilbert (TV show), 131
Dix, Norman, 30
DNA, 126, 131, 133, 140, 166–67
Donohue, Bill, 165–66
doomsday clock, 164
Doppler effect, 72
dualities, 137, 138, 139, 143n2, 188, 193, 194

Einhorn, Martin, 169
Einstein, Albert, 9, 32, 33, 39, 46, 63, 64, 70, 77, 87, 107, 108, 120, 133, 140, 149, 176, 177, 179, 180
Einstein field equations, 39, 75, 175–76, 187, 189
  See also general theory of relativity
Ellis, George, 47, 52, 54, 61
entropy, 56, 67n4, 100–101, 184
  black holes and, 57, 60, 76, 137, 147, 184
Euclidean geometry, 176, 181n2
Euclidean quantum gravity, 92, 121
Euclidean space-time, 89, 90, 146, 153, 154, 176
event horizon. *See* black holes, horizon

false vacuum. *See* inflation
Fella, Judy, 74, 88, 96, 97, 98, 99, 113
Ferguson, Kitty, 113
Ferneyhaugh, Roger, 24
Feynman, Richard, 70, 84, 153, 169
Filkin, David, 124
flat universe, 178, 180, 187

Florides, Petros, 153
Focus (magazine), 140
Freedman, Gordon, 119
Friedmann, Alexander, 176
Friedmann-Robertson-Walker model, 176, 186–87
fundamental forces, 85, 91, 93n14, 102
Futurama (TV show), 131
Future of Theoretical Physics and Cosmology, The (Gibbons, Shellard, and Rankin, eds.), 148

G17. *See* International Conference on General Relativity and Gravitation
Galfard, Christophe, 168–69
Galileo, 21, 73–74, 149, 165
gamma ray astronomy, 64, 69
Gamow, George, 32, 37, 48
Gell-Man, Murray, 98, 100
general theory of relativity, 32, 39, 46, 70, 175
See also special theory of relativity
Genewatch, 140
Gentry, Laura, 96, 98, 99
George's Secret Key to the Universe (Hawking and Hawking), 163–64
Geroch, Robert, 54
Gibbons, Gary, 72, 75, 76, 87, 121, 138, 147
global warming, 164
God and Stephen Hawking (Hawdon), 133
God Created the Integers (Hawking), 163
Gold, Thomas, 37
Goldhaber, Alfred, 129
Gold Star Sardine Bar, 112
Gonville and Caius College, 47, 52, 54, 61
See also Cambridge
Gore, Al, 133
Grand Design, The (movie), 163
grandfather paradox. *See* time travel
grand unified theories, 185–86, 188
See also theory of everything
Graves, Robert, 24
gravitational fields, 46, 56–57, 187
See also black holes; cosmology
Green, Michael, 136
Groening, Matt, 162
Gross, David, 103, 121, 134, 167
Groundhog Day (film), 118
Guazzard, Peter, 107
Guseynov, O. H., 72
Guth, Alan, 84–86, 87, 185, 189, 191

Hartle, James, 57, 65, 70–71, 88, 89, 90, 100, 103, 129, 149, 157, 161
Hawdon, Robin, 133
Hawking (TV show), 152
Hawking, Edward, 25
Hawking, Elaine (Mason), 98, 113–14, 119, 122–23, 125, 148, 168
Hawking, Frank, 21–23, 24–25, 26, 42, 43, 45, 99
Hawking, Isobel, 21–23, 24, 99
Hawking, Jane (Wilde), 24, 38, 54, 69, 73–74, 81–82, 87, 119
autobiography, 123, 132–33, 152
courtship, 40–41, 42–44, 45–46, 48
disabled rights and, 52, 59, 113, 132
divorce, 114, 122, 168
education, 24, 38–39, 45–46, 53, 54, 74–75, 78, 83
effects of Stephen's illness on, 42, 52, 66, 75, 82, 96–97, 98–99, 113, 132
Jonathan Hellyer Jones and, 76, 78, 96, 114, 123, 124, 148
pregnancies, 53, 54–55, 78. *See also* Hawking, Lucy; Hawking, Robert; Hawking, Timothy
religion and, 43, 48, 76, 77, 108, 114

Hawking, Lucy, 9, 55, 56, 66, 75, 77, 82–83, 95, 99, 110, 112, 113, 114, 124, 133, 135, 151–52, 163–64
Hawking, Mary, 22, 23, 41, 53
Hawking, Philippa, 22, 44
Hawking, Robert, 53, 61, 66, 75, 86, 87, 95, 96, 97, 99, 113, 114, 122
Hawking, Stephen
accidents, 34, 117, 148
awards and honors, 48, 52, 54, 57, 73, 75, 78, 87, 112, 119, 132, 134, 146, 151, 161
bets, 72, 73, 115, 125, 130, 150, 153–54, 158
charities and, 78, 119, 124, 135
communication difficulties, 40, 41, 54, 62, 83, 97–98, 99, 107, 131, 148, 167–68
controversy, 71, 81, 86, 101–102, 104, 121–22, 129–30, 136–37, 140, 145, 165, 169
disabled rights and, 52, 59, 114, 123–24, 134–35, 147, 158–59
divorces, 114, 122, 168
dreams, 23, 42, 97, 148, 166, 170
education, 22, 24–25, 26–27, 37–40, 44, 51, 53
employment, 47–48, 52, 54
fame, 59, 96, 112–13, 121, 124, 126, 131, 133–34, 141, 150, 158, 161–63
health problems, 25, 40–42, 43–45, 52, 53, 55, 62, 66, 75, 81, 96–98, 117, 132, 148, 151, 167
marriage to Elaine Mason, 123
marriage to Jane Wilde, 45–46, 48
music and, 31, 42, 43, 44, 59, 96, 121, 123, 131, 135, 148
nuclear disarmament and, 45, 87, 126, 140, 164–65
nursing care and, 8, 53, 81, 83, 97, 98–99, 110, 113–14
publications, 7, 10, 52, 57, 64, 70, 71, 83, 84, 108, 121, 123, 125, 126, 134, 137, 145–47, 148–49, 152, 162, 163, 167
research methods, 57, 65–66, 77, 109, 117, 130, 157
rowing, 30, 31, 34, 124
and television, 75, 131, 133, 134, 148, 150, 152, 162–63, 168, 169
Hawking, Timothy, 78, 123, 135
Hawking Lectures, The (TV show), 148
Hawking Paradox, The (TV show), 168, 169
Hawking radiation. *See* black holes, radiating
Heart of a Dog, The (Bulgakov), 110
Heisenberg uncertainty principle, 63
Hellyer Jones, Jonathan, 76, 78, 96, 114, 123, 124, 148
Hertog, Thomas, 169
Herzog, Chaim, 108
hierarchy problem, 139
Higgs, Peter, 149–50, 188
Higgs field, 188–89
Higgs particle, 92, 94n36, 149–50, 158, 188, 191
Hollywood Reporter (publication), 163
holographic principle, 193
horizon and black holes, 46, 56–57, 60, 63, 76, 137, 153, 183, 184, 189, 193
de Sitter and, 76, 138
Hoyle, Fred, 32, 33, 37, 38, 39, 44, 46, 47, 48, 81
Hubble, Edwin, 32, 87
Hubble's law, 32, 85, 176
Hubble Space Telescope, 123

Illustrated A Brief History of Time, The (Hawking), 123

imaginary numbers, 88–89
imaginary time, 88–90, 109, 111, 117–18, 146, 153, 154, 181n2
"Imagination and Change: Science in the Next Millennium" (Hawking), 125
Independent (newspaper), 165
inflation, 84–88, 92, 138, 139, 146, 176, 178, 179–80
  inflationary cosmology, 84, 85, 87, 170, 185–92
  new inflation, 86, 87, 93n17, 129, 189–91
  open inflation, 129–31
information paradox. *See* black holes
Institute of Astronomy, 53, 57, 60, 61
Institute of Physics, 53, 81, 108
Intel Corporation, 124, 134
International Conference on General Relativity and Gravitation, 153, 168–69
Iraq War, 165
irreducible mass, 56, 57
Israel, Werner, 65, 77, 108
"Is the End in Sight for Theoretical Physics?" (Hawking), 70, 83

Jaded (Hawking, Lucy), 151
John D. and Catherine T. MacArthur Foundation, 98, 100
John Paul II (pope), 74, 84, 100, 165

Kaku, Michio, 70
Kalunza, Theodore, 136
Kane, Gordon, 104, 150
"Keep Talking" (song), 121
Kepler, Johannes, 149
Key to the Universe, The (TV show), 75
King, Basil, 24, 40
King, Diana, 38, 40, 42
King, Larry, 133
Kings College, 40, 47, 54, 65
KipFest Saturday Science Talks, 134
Klein, Oscar, 136

La Flamme, Raymond, 101
Lapades, Allan, 75
Large Electron Positron, 150
Late Night with Conan O'Brien (TV show), 150
Lemaître, Georges, 32, 83, 176
LEP. *See* Large Electron Positron
Linde, Andrei, 86, 87, 93n17, 130, 190
London Mathematical Society, 132
London Telegraph (newspaper), 130
London University, 53
Lost (TV show), 162–63
Lou Gehrig's disease. *See* ALS (Amyotrophic lateral sclerosis)
Lucas, Henry, 78

Macmillan Publishing, 123
Manchester Guardian (newspaper), 130
Martin, Philip, 152
Mason, David, 99, 123
Mason, Elaine. *See* Hawking, Elaine (Mason)
Masters of Science Fiction (TV show), 163
Mayer, Sue, 140
McClenahan, John, 24, 25, 53
Microsoft, 122
Mitchell, John, 46
Mlodinow, Leonard, 163
Moore, Gordon, 134
Morris, Errol, 119
Moss, Ian, 87
motor neuron disease. *See* ALS (Amyotrophic lateral sclerosis)
Motor Neuron Disease Association, 78, 147

"Moulin, The," 111, 123
M-theory, 137, 138, 139, 149, 153, 159
  See also branes
Murdin, Paul, 72
Murray, Bill, 118
Music to Move the Stars (Hawking, Jane). *See* Hawking, Jane (Wilde), autobiography
My Glorious Breakdown (Hawking, Lucy), 152

naked singularity. *See* cosmic censorship conjecture
Narlikar, Jayant, 38, 39, 44, 46, 48
Nature (journal), 65, 72, 108, 163
New Millennium Press, 148
Newton, Isaac, 53, 78, 108, 120, 133, 140, 149
New York Times (newspaper), 109
Nichols, Nichelle, 131
Nimoy, Leonard, 120
Nobel Prize, 157
no-boundary proposal, 84, 88, 89, 90, 100, 101, 110, 111, 129–30, 139, 163, 165, 169
no hair theorem. *See* black holes, uniqueness theorems and
Nuffield Workshop on the early universe, 87–88, 92, 191

OK Computer (Radiohead), 131
On the Shoulders of Giants (Hawking), 149, 163
open universe, 130, 178
origin of universe. *See* big bang; inflation; no-boundary proposal
Overbye, Dennis, 77
Oxford, 21–22, 26, 27, 29–34, 38–39, 40, 42, 65, 77, 110, 113, 114, 124, 161

Page, Don, 69, 76, 101–102, 137–38
particle physics, 32, 69, 84, 85, 98, 136, 152–53
  See also quantum mechanics
path integrals. *See* sum over histories method
Paul Dirac Centennial Celebration, 149
Pauli exclusion principle, 91
Paul VI (pope), 73
P-branes. *See* branes
Peiser, Benny, 140
Penrose, Roger, 47, 52, 54, 55, 56–57, 62, 64–65, 73, 108, 121–22, 125, 136, 147–48
Penzias, Arno, 48
People (magazine), 112
Persaud, Ray, 146
phase transitions, 85, 189, 191
Phillips, Nick, 113
Physical Review (journal), 77
Physical Review D (journal), 169
PhysicsWeb, 140
Physics World (journal), 132, 140
Pink Floyd (music group), 121, 131, 135
Playboy (magazine), 112
Polchinski, Joe, 137
Powney, Derek, 30, 31
Preskill, John, 125, 153–54
Pride Mobility (company), 134
Principia Mathematica (Newton), 108
Proceedings of the Royal Society, 54

quantum gravity, 61, 65, 70, 76, 85, 86, 92, 121–22, 137–38, 153, 194
quantum mechanics, 32, 63, 88, 90, 91, 118, 121, 137, 146, 188, 194
  black holes and, 62–63, 65, 71, 75–76, 137, 183

general relativity and, 61–62, 64, 70, 76, 83, 84, 149
quantum tunneling, 189–90

radio astronomy, 38
Radiohead (music group), 131
Randall, Lisa, 139
Real Stephen Hawking, The (TV show), 134
Rees, Martin, 64, 81, 98, 148
religion, 24, 48, 76, 78, 104, 108, 109–10, 114, 130, 165
See also atheism; Vatican
Ross, Jonathan, 162
Royal Astronomical Society, 73, 96
Royal Society, 44, 46, 48, 54, 66, 75, 109, 161, 164
Run for Your Life (Hawking, Lucy), 164
Ryle, Martin, 38, 44, 48

Sagan, Carl, 9, 67, 109–10, 117, 126
scalar field, 188, 191
Schwarz, John, 136
Schwarzchild, Karl, 46
Schwarzschild radius. *See* horizon and black holes
Sciama, Dennis, 37–38, 39–40, 43, 47–48, 52, 53, 54, 61, 64, 65, 100, 161
Science (journal), 130
Sellers, Pier, 161–62
Silicon Graphics Inc., 124, 152
Silk, Joseph, 146
Simpsons, The (TV show), 131, 133, 162
"Singularities and the Geometry of Space-Time" (Hawking), 52
singularity, 46, 57, 71, 88, 89, 125
big bang and, 54, 55
black hole and, 11, 54, 55
naked singularity, 125, 184. *See also* cosmic censorship conjecture
theorems and, 47, 54, 73, 100, 152, 159
"Sixty Years in a Nutshell" (Hawking), 147
Smith, Alex Mackenzie, 124, 151
Smith, Diane, 159
special theory of relativity, 63, 152
See also general theory of relativity
Spielberg, Steven, 119
Spiner, Brent, 120
St. Albans, 23, 24–25, 27, 34, 38, 44
standard model. *See* particle physics
Starobinsky, Alexei, 62
Star Trek, 111, 118, 120, 131, 159
Status Quo (music group), 135
steady state model, 32, 33, 37–38, 44, 46, 48
Steinhardt, Paul, 87, 93n17, 190
Stephen Hawking's Beyond the Horizon (film), 152
Stephen Hawking's Universe (TV series), 124, 152
string theory, 121, 134, 136–37, 138–39, 153, 169, 193, 194
Strominger, Andrew, 137
Study Week on Cosmology and Fundamental Physics (Vatican), 83, 165
sum over histories method, 70, 84, 88, 109, 153, 169
Sunday Times (newspaper), 109
Sundrum, Raman, 139
supercooling, 85, 189, 190
supergravity, 90, 92, 136, 137, 138, 139, 194
superstring. *See* string theory
supersymmetry, 90–92, 136, 137, 138, 194
Susskind, Leonard, 135, 136, 158, 169
symmetry breaking, 187–89, 192

Taylor, John G., 65
television and Stephen Hawking, 75, 131, 133, 134, 148, 150, 152, 162–63, 168, 169
Terminal Error (film), 7
Texas Symposium on Relativistic Astrophysics, 8, 57, 102
theory of everything, 61, 85, 112, 136, 148–49
Theory of Everything, The (Hawking), 148–49
thermodynamics, 75, 100–101, 102, 103, 138
  black holes and, 55–56, 60–61, 63, 76, 183–84
't Hooft, Gerard, 154, 193
Thorne, Carolee, 125
Thorne, Kip, 48, 57, 62, 65, 66, 72, 98, 117–19, 134
  bets, 72, 73, 115, 125, 153–54
time travel, 118–19, 123
Trimble, Virginia, 72
true vacuum. *See* inflation
Turner, Michael, 87
Turok, Neil, 129–31
Tutu, Desmond, 134
TV, 7

U2 (music group), 148
unified theories. *See* theory of everything
uniqueness theorems. *See* black holes, uniqueness theorems and
Unity of the Universe, The (Sciama), 37
Universe in a Nutshell, The (Hawking), 145–47, 152
Unruh, William, 154

vacuum energy, 79n21, 176, 179
Vafa, Cumrun, 137
Vatican, 73–74, 84, 100, 165, 183
Vilenkin, Alexander, 130
Virgin Galactic (company), 166
virtual particles, 63, 118, 150

Wagner, Richard, 42, 43, 44, 96, 123
Warner, Nick, 92
Webster, B. Louise, 72
weightless flight, 166
Wheeler, John Archibald, 55, 56, 60, 65, 119
Whiting, Bernard, 83
"Why I Have Not Changed My Mind" (Hawking), 137
Wilde, Jane. *See* Hawking, Jane (Wilde)
Wilkinson, Jonny, 162
Wilkinson Mapping Anisotrophy Probe, 178–79, 180, 181n3, 191
Wilson, Robert, 48
Witten, Edward, 137
WMAP. *See* Wilkinson Mapping Anisotrophy Probe
Woltosz, Walt, 98
Woolley, Richard, 32
Words + (company), 168
wormhole, 110–11, 117–19, 123

X Prize Foundation, 166
x-ray astronomy, 72

Zeldovich, Yakov, 62, 65, 72
Zero Gravity (company), 166
Zuckerman, Al, 95